Communications
in Computer and Information Science 888

Commenced Publication in 2007
Founding and Former Series Editors:
Phoebe Chen, Alfredo Cuzzocrea, Xiaoyong Du, Orhun Kara, Ting Liu,
Dominik Ślęzak, and Xiaokang Yang

Editorial Board

More information about this series at http://www.springer.com/series/7899

Zhi-Hua Zhou · Qiang Yang
Yang Gao · Yu Zheng (Eds.)

Artificial Intelligence

First CCF International Conference, ICAI 2018
Jinan, China, August 9–10, 2018
Proceedings

 Springer

Editors
Zhi-Hua Zhou
Nanjing University
Nanjing
China

Qiang Yang
Hong Kong University of Science
  and Technology
Hong Kong SAR
China

Yang Gao
Nanjing University
Nanjing
China

Yu Zheng
JD Finance
Beijing
China

ISSN 1865-0929   ISSN 1865-0937 (electronic)
Communications in Computer and Information Science
ISBN 978-981-13-2121-4   ISBN 978-981-13-2122-1 (eBook)
https://doi.org/10.1007/978-981-13-2122-1

Library of Congress Control Number: 2018950655

This Springer imprint is published by the registered company Springer Nature Singapore Pte Ltd.
The registered company address is: 152 Beach Road, #21-01/04 Gateway East, Singapore 189721, Singapore

Preface

Welcome to the First CCF International Conference on Artificial Intelligence (CCF-ICAI 2018) held in Jinan, China!

CCF-ICAI, hosted by the China Computer Federation (CCF) and co-organized by CCF Artificial Intelligence & Pattern Recognition Society, is an international conference in the field of artificial intelligence. It aims to provide a leading international forum for researchers, practitioners, and other potential users in artificial intelligence and related fields to share their new ideas, progress, and achievements. The selected papers included in the proceedings not only address challenging issues in various aspects of artificial intelligence, but also synthesize contributions from related disciplines that illuminate the state of the art.

CCF-ICAI 2018 received 82 full submissions. Each submission was reviewed by at least three reviewers. Based on the reviewers' comments, 17 papers were finally accepted for presentation at the conference, with an acceptance rate of 20.7%. The proceedings of CCF-ICAI 2018 are published as a dedicated volume of CCIS by Springer.

We are grateful to Prof. Masashi Sugiyama from RIKEN/University of Tokyo, Dr. Ming Zhou from Microsoft Research Asia, and Prof. Shuicheng Yan from National University of Singapore/Qihoo 360 AI Institute for giving keynote speeches at CCF-ICAI 2018.

The high-quality program would not have been possible without the authors who chose CCF-ICAI 2018 as a venue for their publications. We are also very grateful to the Organizing Committee members, the Program Committee members, and the reviewers, who made a tremendous effort in soliciting and selecting research papers with a balance of high quality, new ideas, and novel applications. We appreciate Springer for publishing the conference proceedings, and we are particularly thankful to Celine Chang, Jane Li, and Leonie Kunz from Springer for their efforts and patience in collecting and editing these proceedings.

We hope you find the proceedings of CCF-ICAI 2018 an enjoyable and fruitful reading experience!

July 2018

Zhi-Hua Zhou
Qiang Yang
Yang Gao
Yu Zheng

Organization

General Chairs

Zhi-Hua Zhou	Nanjing University, China
Qiang Yang	Hong Kong University of Science and Technology, Hong Kong SAR, China

Program Chairs

Yang Gao	Nanjing University, China
Yu Zheng	JD Finance, China

Publication Chairs

Min-Ling Zhang	Southeast University, China
Chaoran Cui	Shandong University of Finance and Economics, China

Local Arrangements Chairs

Peiguang Lin	Shandong University of Finance and Economics, China
Xiushan Nie	Shandong University of Finance and Economics, China

Publicity Chair

Muwei Jian	Shandong University of Finance and Economics, China

Finance Chair

Shanshan Gao	Shandong University of Finance and Economics, China

Program Committee

En-Hong Chen	University of Science and Technology of China, China
Songcan Chen	Nanjing University of Aeronautics and Astronautics, China
Chaoran Cui	Shandong University of Finance and Economics, China
Shifei Ding	China University of Mining and Technology, China
Jun Dong	Chinese Academy of Sciences, China
Hongbin Dong	Harbin Engineering University, China
Junping Du	Beijing University of Posts and Telecommunications, China
Jufu Feng	Peking University, China
Yang Gao	Nanjing University, China
Xin Geng	Southeast University, China

Maozu Guo	Beijing University of Civil Engineering and Architecture, China
Zhongshi He	Chongqing University, China
Qing He	Chinese Academy of Sciences, China
Qinghua Hu	Tianjin University, China
Jianbin Huang	Xidian University, China
Genlin Ji	Nanjing Normal University, China
Muwei Jian	Shandong University of Finance and Economics, China
Liping Jing	Beijing Jiaotong University, China
Yujian Li	Beijing University of Technology, China
Ming Li	Nanjing University, China
Yu-Feng Li	Nanjing University, China
Qingyong Li	Beijing Jiaotong University, China
Wu-Jun Li	Nanjing University, China
Fanzhang Li	Soochow University, China
Jiye Liang	Shanxi University, China
Xu-Ying Liu	Southeast University, China
Xinjun Mao	National University of Defense Technology, China
Qiguang Miao	Xidian University, China
Xiushan Nie	Shandong University of Finance and Economics, China
Dantong Ouyang	Jilin University, China
Yuhua Qian	Shanxi University, China
Lin Shang	Nanjing University, China
Zhenwei Shi	Beihang University, China
Zhong Su	IBM China Research Lab, China
Jiankai Sun	The Ohio State University, USA
Qing Tao	Army Officer Academy of PLA, China
Jianhua Tao	Chinese Academy of Sciences, China
Xiangrong Tong	Yantai University, China
Wenjian Wang	Shanxi University, China
Li Wang	Taiyuan University of Technology, China
Hongyuan Wang	Changzhou University, China
Guoyin Wang	Chongqing University of Posts and Telecommunications, China
Yunhong Wang	Beihang University, China
Yimin Wen	Guilin University of Electronic Technology, China
Jianxin Wu	Nanjing University, China
Juanying Xie	Shaanxi Normal University, China
Yong Xu	South China University of Technology, China
Xinshun Xu	Shandong University, China
Yubin Yang	Nanjing University, China
Bo Yang	Jilin University, China
Yan Yang	Southwest Jiaotong University, China
Ming Yang	Nanjing Normal University, China
Qiang Yang	Hong Kong University of Science and Technology, Hong Kong SAR, China

Minghao Yin	Northeast Normal University, China
Yilong Yin	Shandong University, China
Zhiwen Yu	South China University of Technology, China
Jian Yu	Beijing Jiaotong University, China
Daoqiang Zhang	Nanjing University of Aeronautics and Astronautics, China
Li Zhang	Soochow University, China
Zhao Zhang	Soochow University, China
Min-Ling Zhang	Southeast University, China
Changshui Zhang	Tsinghua University, China
Yi Zhang	Sichuan University, China
Yu Zheng	JD Finance, China
Yong Zhou	China University of Mining and Technology, China
Zhi-Hua Zhou	Nanjing University, China
Fuzhen Zhuang	Chinese Academy of Sciences, China
Quan Zou	Tianjin University, China

Contents

Unsupervised Learning

An Improved DBSCAN Algorithm Using Local Parameters

Kejing Diao[1], Yongquan Liang[1,2,3(✉)], and Jiancong Fan[1,3]

[1] College of Computer Science and Engineering,
Shandong University of Science and Technology, Qingdao, China
lyq@sdust.edu.cn
[2] Provincial Key Laboratory for Information Technology of Wisdom Mining
of Shandong Province, Shandong University of Science and Technology,
Qingdao, China
[3] Provincial Experimental Teaching Demonstration Center of Computer,
Shandong University of Science and Technology, Qingdao, China

Abstract. Density-Based Spatial Clustering of Applications with Noise (DBSCAN), as one of the classic density-based clustering algorithms, has the advantage of identifying clusters with different shapes, and it has been widely used in clustering analysis. Due to the DBSCAN algorithm using globally unique parameters ε and MinPts, the correct number of classes can not be obtained when clustering the unbalanced data, consequently, the clustering effect is not satisfactory. To solve this problem, this paper proposes a clustering algorithm LP-DBSCAN which uses local parameters for unbalanced data. The algorithm divides the data set into multiple data regions by DPC algorithm. And the size and shape of each data region depends on the density characteristics of the sample. Then for each data region, set the appropriate parameters for local clustering, and finally merge the data regions. The algorithm is simple and easy to implement. The experimental results show that this algorithm can solve the problems of DBSCAN algorithm and can deal with arbitrary shape data and unbalanced data. Especially in dealing with unbalanced data, the clustering effect is obviously better than other algorithms.

Keywords: Clustering · Unbalanced data · Local parameters

1 Introduction

The research of clustering algorithm [1–3] plays an important role in the field of data mining. It has an extensive application [4, 5] in pattern recognition, decision-making, biological and commercial research, and so on. There are four kinds of methods commonly used in cluster analysis: division-based clustering method, level-based clustering method, density-based clustering method and grid-based clustering method. The Density-Based Spatial Clustering of Applications with Noise [6] is a typical density-based clustering algorithm, which has the advantages of insensitive to noise points and can find clusters of arbitrary shape and size. When clustering unbalanced data, the result of clustering is inaccurate, because the DBSCAN algorithm uses globally unique parameters ε and MinPts [7]. In response to the above problems, some

© Springer Nature Singapore Pte Ltd. 2018
Z.-H. Zhou et al. (Eds.): ICAI 2018, CCIS 888, pp. 3–12, 2018.
https://doi.org/10.1007/978-981-13-2122-1_1

scholars have conducted some research. Literature [8] proposed a PDBSCAN algorithm based on data partitioning, using manual methods to achieve partitioning. In
order to improve the efficiency of DBSCAN algorithm, Literature [8] proposed a new
parallel algorithm using disjoint-set data structure for clustering. In [10], the DN-
DBSCAN algorithm based on dynamic neighbors is proposed to solve the problem that
the parameters are not easily accessible.

In this paper, we propose a clustering algorithm LP-DBSCAN with local parameters. Using the DPC algorithm [11], we divide the sample into several small data sets.
Then according to the density of each small data set, set parameters for clustering.
Finally, merge all the small data sets. The traditional data partitioning [12–14] divides
the data sets into some regular rectangles, which is not applicable to the spiral class.
This paper divides the data sets into irregular shapes and can be applied to arbitrary
shape classes. The experiment proved that the algorithm is simple and easy to
implement, and it is suitable for arbitrary shape data and unbalanced data. In dealing
with unbalanced data, the clustering effect has been significantly improved.

2 DBSCAN Algorithm

The DBSCAN algorithm is a classic density-based clustering algorithm that can rec-
ognize noise and find classes of any shape and size. DBSCAN can define the class as
the set of sample points connected by the maximum density-connected from the
density-reachable relationship. The core idea is to find out the core objects, connect the
core objects and their neighborhoods, and form densely connected areas as clusters.

However, there are some problems with the DBSCAN algorithm. DBSCAN
requires a lot of memory and I/O overhead when working with large data sets.
Since DBSCAN uses global uniform parameters, it is difficult to achieve ideal clustering when dealing with unbalanced data. If the neighborhood radius ε is set too small,
it will lead to the sparse classes being divided into adjacent classes or treated as noise
points. If the neighborhood radius ε is set too large, the classes with smaller distance
between classes or the more intensive distribution can be converged into one category.
As is shown in Fig. 1, if the neighborhood radius ε is set too small, the C4 part of the
point will be treated as a noise point, and the more concentrated parts of C4 will be
converged into one kind. If the neighborhood radius ε is set too large, C1 and C2 will
be classified into one category, C8, C9, and C10 will be classified into one category,
C11, C12 and C13 will be classified into a category.

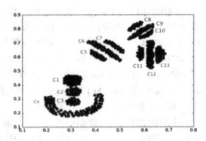

Fig. 1. Unbalanced data. There are 13 classes, which are represented by C1 to C13.

3 LP-DBSCAN Algorithm

3.1 Basic Concepts

Aiming at a series of problems caused by using a globally unique parameter for the above mentioned DBSCAN algorithm, we proposed a clustering algorithm LP-DBSCAN using local parameters. The basic idea of the algorithm is as follows: Firstly, divided the sample into several irregular data regions $C(C_1, C_2, \cdots, C_n)$ according to the density trend of the samples by DPC algorithm. The size and shape of each data area depend on the trend of the density of the original sample, not the regular shape. Then set the appropriate parameters for each data area $C_i(i = 1, 2, \cdots, n)$ according to the density features of each data area $C(C_1, C_2, \cdots, C_n)$. Finally, merge the divided data regions $C(C_1, C_2, \cdots, C_n)$. When merging, it is necessary to determine whether the classes in the adjacent data regions need to be merged according to the merger condition, and need to reclassify the noise points.

The DPC algorithm divides the data sets according to the peak density and distance. Therefore, for some unbalanced data sets, DPC algorithm can divides classes with large differences in density into different data regions. At this point, the density of the classes in each data regions is not much different. Then using DBSCAN algorithm to use local parameters for clustering can achieve good results.

3.2 Merger Conditions

In order to illustrate the conditions of the merger, the following definitions need to be quoted.

Definition 1. Boundary point: A boundary point is any point in the data sample, itself is not a core point, but it is in the neighborhood of a core data point.

Definition 2. Noise Point: Any point in the data sample that is neither a core point nor a boundary point.

Merger conditions can be divided into strong merger conditions and weak merger conditions, the use of which conditions for the merger of the adjacent neighborhoods may be based on the actual situation.

Weak merger conditions: Two classes A and B are merged if and only if: ① A and B are in two adjacent data regions P_A, P_B; ② $\exists p \in E_A, \exists q \in E_B$, satisfies DISTANCE$\{p, q\} \leq \varepsilon(P_A, P_B)$. ($E_A, E_B$ are the boundary point set of class A and class B, P_A, P_B are the neighborhood radius of class A and class B, $\varepsilon(P_A, P_B)$ is the smaller neighborhood radius in partition P_A, P_B).

Strong merger conditions: Two classes A and B are merged if and only if: ① A and B are in two adjacent data regions P_A, P_B; ② $\forall p_i \in E_A, \forall p_j \in E_B, N_A = |E_A|, N_B = |E_B|$, satisfies

$$\frac{\sum_i \sum_i DISTANCE(p_i, p_j)}{N_A \times N_B} \leq \varepsilon(P_A, P_B) \tag{1}$$

3.3 Noise Point Processing

When dividing a data set into multiple data regions, it is possible to divide a small class into multiple partitions, which can cause the points in the class to be treated as noise points when performing local clustering in the partitions. Therefore, after the merged partition is completed, we need to reprocess the noise point. The processing of the noise point can be divided into two steps, firstly, judge whether the current noise point can be merged into the existing class; secondly, judge whether the current noise point can be aggregated into a new class.

Noise Points are Merged into Existing Classes

A noise point p belongs to class C if and only if: ① p and C are in two adjacent data regions P_A, P_B; ② $\exists q \in E_C$, satisfies DISTANCE$\{p, q\} \leq \varepsilon(P_C)$. Similarly, this condition can be divided into strong merger conditions and weak merger conditions. The above conditions is weak merger conditions. Strong merger conditions only need to strengthen condition ② in weak merge conditions as $\forall q_i \in E_C, N_C = E_C$ satisfies

$$\frac{\sum_i DISTANCE(p, q_i)}{N_C} \leq \varepsilon(P_C) \tag{2}$$

Generate a new class from the noise point if and only if: $P_A, P_B, \cdots, P_N$ are adjacent data partitions, S is a set of noise points in the $P_A, P_B, \cdots, P_N$ data areas, and $\exists p_1, \exists p_2, \cdots \exists p_i \in S (i \geq MinPts)$, satisfies DISTANCE$\{p_1, p_i\} \leq \varepsilon(P_A, P_B, \cdots, P_N)$, then p_1 is a new core point in the global, $\{p_i\} = (1, 2, \cdots, m)(m \geq MinPts)$ is the point within this new class.

3.4 Time Complexity Analysis

The algorithm is mainly composed of the following three parts: use the DPC algorithm to divide the dataset, the time complexity is $O(n^2)$; use the local parameters of DBSCAN algorithm to clustering the data regions, the time complexity is $O(n \log n)$, according to the class merging conditions, and process the noise points with a time complexity of $O(n^2)$, when using R*-tree regional query, the time complexity is $O(n \log n)$.

4 Results and Experimental Comparative Analysis

In order to test the clustering effect of the algorithm, the experiment is divided into two parts: the artificial data set and the real data set. The real data set comes from the UCI machine learning database. Comparison algorithm uses DPC algorithm, DBSCAN algorithm and K-means algorithm [15]. Among them, DPC algorithm is a typical density peak algorithm, and K-means is a typical algorithm based on partition. The structure of artificial data set and UCI real data set is shown in Table 2.

Table 1. Artificial data sets and real data sets used in the experiment

Data set	Number	Attributes	Class
Flame	240	2	2
Dpc	1000	2	6
Ds3c3sc6	905	2	6
Ds2c2sc13	588	2	13
Iris	150	4	3
Wine	178	13	3
Seeds	210	7	3
Art	300	4	3
Glass	214	9	6
Ionosphere	351	34	2
Ecoli	336	7	8
Segmentation	2310	11	7
Soybeansmall	47	35	4
Spiral3	240	2	2

The first four data sets in Table 1 are artificial data sets and the last ten data sets are UCI real data sets. These data sets are commonly used to test the performance of clustering algorithm, they have a great difference in the sample size, the number of attributes and the number of clusters. The evaluation index of clustering algorithm adopts three kinds of evaluation indexes: Accuracy, recall and NMI. The range of the three evaluation indexes is [0, 1], and the larger the value is, the better the clustering effect will be. The definitions are as follows:

$$Accuracy = \frac{TP}{TP + FP} \tag{3}$$

$$Recall = \frac{TP}{TP + FN} \tag{4}$$

TP (True Positive) refers to the positive tuple correctly classified by the classifier, FP (False Positive) refers to the negative tuple that the classifier incorrectly marks as the positive tuple, FN (False Negative) refers to the positive tuple that the classifier incorrectly marks as the negative tuple.

$$NMI = \frac{2 \times [H(A) + H(B) - H(A, B)]}{H(A) + H(B)} \tag{5}$$

P_A, P_B represent the probability distribution of A, B, P_{AB} represents the joint probability distribution of A and B, $H(A) = -\sum_a P_A(a) \log P_A(a)$, $H(B) = -\sum_b P_B(b) \log P_B(b)$, $H(A, B) = -\sum_{a,b} P_{AB}(a, b) \log P_{AB}(a, b)$.

4.1 Artificial Data Set Clustering Results and Analysis

In order to facilitate visualization, the four artificial data sets used in this paper are two-dimensional data sets. In order to test and demonstrate the clustering effect of the LP-DBSCAN algorithm on the balanced data set and the unbalanced data set, we choose two balanced data sets and two unbalanced data sets, and the clustering results of the algorithm on artificial data sets are compared with the DBSCAN algorithm. We use graphs to show the clustering results, with different colors and shapes representing different clusters, with black solid dots representing noise points.

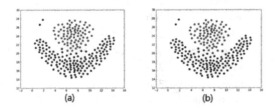

Fig. 2. Flame data set (a) clustering results of the LP-DBSCAN algorithm on flame data set. (b) clustering results of the DBSCAN algorithm on flame data set.

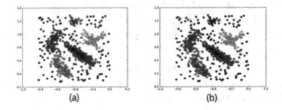

Fig. 3. Dpc data set (a) Clustering results of the LP-DBSCAN algorithm on dpc data set. (b) Clustering results of the DBSCAN algorithm on dpc data set.

As is shown in the Figs. 2 and 3, LP-DBSCAN algorithm has better clustering result than DBSCAN algorithm in the classes with similar data densities. The clustering result of the LP-DBSCAN algorithm is slightly better than that of the DBSCAN algorithm. The LP-DBSCAN algorithm can find classes of any shape and size and it is insensitive to noise points. For unbalanced data sets, that is, data sets with large data densities and the inhomogeneous distance between clusters, the clustering result of the LP-DBSCAN algorithm in this paper is better than that of the DBSCAN algorithm.

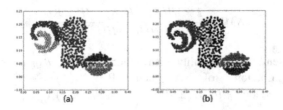

Fig. 4. Ds3c3sc6 data set (a) Clustering results of the LP-DBSCAN algorithm on ds3c3sc6 data set. (b) Clustering results of the DBSCAN algorithm on ds3c3sc6 data set.

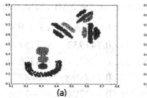

Fig. 5. Ds3c3sc13 data set (a) Clustering results of the LP-DBSCAN algorithm on ds3c3sc13 data set. (b) Clustering results of the DBSCAN algorithm on ds3c3sc13 data set.

As is shown in the Figs. 4 and 5, there are 6 classes in the ds3c3sc6 data set, 6 classes can be found in the LP-DBSCAN algorithm, the clustering accuracy rate is 0.719337, while DBSCAN algorithm can only find 5 classes, the clustering accuracy rate is 0.539227. There are 13 classes in the ds3c3sc13 data set, 11 classes can be found by the LP-DBSCAN algorithm, the clustering accuracy rate is 0.874150, while the DBSCAN algorithm can only find 8 classes, the clustering accuracy rate is 0.731293. Due to the DPC algorithm used in this paper, DPC algorithm can not partition some clusters with different densities into different partitions on some data sets, which leads to that in some data sets, as a result, all the classes cannot be found in some data sets, such as Fig. 5(a). However, the proposed LP-DBSCAN algorithm can find more classes than the DBSCAN algorithm, which can be applied to unbalanced data. It can be seen from the experiment that LP-DBSCAN algorithm proposed in this paper is better than DBSCAN on the balanced data set. On unbalanced data sets, the LP-DBSCAN algorithm can find more classes by using local neighborhood parameters (ε, MinPts) and the clustering effect is better.

4.2 Real Data Set Clustering Results and Analysis

The experimental results in this paper are shown in Tables 2, 3 and 4. Table 2 shows the accuracy of each clustering algorithm. Table 3 shows the mutual information of each clustering algorithm. Table 4 shows the recall of each clustering algorithm. The rough data in the table is the best result for each data set.

Table 2. Clustering algorithm on real data set in the ACC evaluation index results

Data set	DBSCAN	DPC	K-MEANS	LP-DBSCAN
Iris	0.680000	0.740000	0.821333	**0.873333**
Wine	0.606742	0.752809	0.642697	**0.880254**
Seeds	0.676190	**0.895238**	0.892857	**0.895238**
Art	0.666667	0.893333	0.852333	**0.923333**
Glass	0.462617	0.537383	0.461682	**0.612150**
Ionosphere	0.646724	0.680912	0.712251	**0.911681**
Ecoli	0.428571	0.726190	0.507738	**0.761905**
Segmentation	0.374026	0.429004	**0.500390**	0.483117
Soybeansmall	0.787234	0.702128	0.704255	**0.936170**
Zoo	0.782178	0.722772	0.697030	**0.871287**

Table 3. Clustering algorithm on real data set in the NMI evaluation index results

Data set	DBSCAN	DPC	K-MEANS	LP-DBSCAN
Iris	0.722114	0.708717	**0.741932**	0.715114
Wine	0.412104	**0.452720**	0.428757	0.438654
Seeds	0.548902	0.694248	**0.71.0068**	0.648731
Art	0.761170	0.746236	0.733794	**0.807755**
Glass	**0.488373**	0.275908	0.387323	0.441214
Ionosphere	0.545751	0.095944	0.387323	**0.556173**
Ecoli	0.195643	0.594806	0.624478	**0.681127**
Segmentation	0.400340	0.396151	**0.580158**	0.447626
Soybeansmall	0.848926	0.621166	0.563094	**0.856399**
Zoo	0.800801	0.623930	0.612205	**0.829422**

Table 4. Clustering algorithm on real data set in the REC evaluation index results

Data set	DBSCAN	DPC	K-MEANS	LP-DBSCAN
Iris	0.680000	0.740000	0.868667	**0.880254**
Wine	0.658333	0.833333	0.694444	**0.877551**
Seeds	0.647887	0.901408	0.845070	**0.929577**
Art	0.666667	0.893333	0.893333	**0.923333**
Glass	0.46729	0.537383	0.542056	**0.612150**
Ionosphere	0.646724	0.668142	0.712251	**0.960177**
Ecoli	0.428571	0.726190	0.657738	**0.761905**
Segmentation	0.325541	0.405195	**0.532900**	0.483117
Soybeansmall	0.787234	0.702128	0.659574	**0.936170**
Zoo	0.782178	0.722772	0.683168	**0.871287**

It can be seen from the data in Tables 2, 3 and 4 that the LP-DBSCAN algorithm proposed in this paper is inferior to the DPC algorithm for the Segmentations data set under the ACC and REC evaluation indexes. For the Seeds data set, the LP-DBSCAN algorithm and DPC algorithm have the same clustering result under the ACC evaluation index and the LP-DBSCAN algorithm is better than the DPC algorithm under the REC index. The LP-DBSCAN algorithm for Iris, Wine, Art, Glass, Ionosphere, Ecoli, Soybeansmall, and Zoo data sets achieved the best results under the ACC and REC evaluation criteria. Under the NMI criteria, the LP-DBSCAN algorithm achieved the best results in the Art, Ionosphere, Ecoli, Soybeansmall and Zoo data sets. For Wine, Seeds and Segmentations data sets LP-DBSCAN algorithm is superior to DBSCAN algorithm under the NMI evaluation index. The clustering results of UCI real data set show that the proposed algorithm has a better clustering effect. Compared with DPC, DBSCAN and K-means algorithms, LP-DBSCAN algorithm can get better clustering results and have stronger robustness.

5 Summary

In this paper, a clustering algorithm named LP-DBSCAN, which is divided by DPC algorithm and using local parameters, is proposed. The algorithm uses DPC algorithm to divide the data set into regions with different shapes according to the trend of sample density. Then, for each region, set the appropriate neighborhood parameters according to the density, that is, using local neighborhood parameters for clustering, which avoids the defects brought by the global parameters of the DBSCAN algorithm. The experimental results of artificial data sets and real data sets show that the LP-DBSCAN algorithm proposed in this paper has better performance and can find clusters of arbitrary shape and size. For the unbalanced data sets, the LP-DBSCAN algorithm can find more numbers clusters, the clustering effect is superior.

The next step will research the automatic determination of local parameters. Since DPC algorithm is adopted to partition in this paper, the parameters are not easy to adjust, and each neighborhood needs its neighborhood parameters ε and MinPts for local clustering. In this paper, the LP-DBSCAN algorithm calculates the distance between points and points in the data set when partitioning, so the next step is to determine the neighborhood parameters ε and MinPts automatically by the distance between points in each partition and the number of points in the partition.

Acknowledgements. As time goes by, I have spent most of my postgraduate studies. Firstly I must thank my mentor for guiding me to get started and motivating me to move forward. Thank him for giving me help in the research ideas of the paper. Secondly, I will thank to the professional brothers in the lab for helping me to revise my paper and for giving me guidance on experiments. Finally, I will sincerely thank to all the experts and professors who help me review the manuscript during the busy schedule.

References

1. Sun, J., Liu, J., Zhao, L.: Clustering algorithms research. J. Softw. **19**(1), 48–61 (2008)
2. Agarwal, S.: Data mining: data mining concepts and techniques. In: International Conference on Machine Intelligence and Research Advancement, pp. 203–207. IEEE (2014)
3. Witten, I.H., Frank, E.: Data mining: practical machine learning tools and techniques. ACM SIGMOD Rec. **31**(1), 76–77 (2011)
4. Hu, Z., Hongye, T., Yuhua, Q.: Chinese text deception detection based on ensemble learning. J. Comput. Res. Dev. **52**(5), 1005–1013 (2015)
5. Chu, X.: K-means clustering algorithm and artificial fish swam algorithm applied in image segmentation technology. Comput. Syst. Appl. **22**(4), 92–94 (2013)
6. Ester, M., Kriegel, H., Sander, J., et al.: A density-based algorithm for discovering clusters in large spatial databases (1996)
7. Xiong, Z., Sun, S., Zhang, Y.: Partition-based DBSCAN algorithm with different parameter. Comput. Eng. Des. **26**(9), 2319–2321 (2005)
8. Zhou, S., Zhou, A., Cao, J.: A data-partitioning-based DBSCAN algorithm. J. Comput. Res. Dev. **37**(10), 1153–1159 (2000)
9. Patwary, M.A., Liao, W., Manne, F., et al.: A new scalable parallel DBSCAN algorithm using the disjoint-set data structure, pp. 1–11 (2012)

10. Li, Y., Ma, L., Fan, F.: Improved DBSCAN clustering algorithm based on dynamic neighbor. Comput. Eng. Appl. **52**(20), 80–85 (2016)
11. Rodriguez, A., Laio, A.: Machine learning. Clustering by fast search and find of density peaks. Science **344**(6191), 1492 (2014)
12. Dai, B.R., Lin, I.C.: Efficient map/reduce-based DBSCAN algorithm with optimized data partition. In: IEEE, International Conference on Cloud Computing, pp. 59–66. IEEE (2012)
13. Sancho-Asensio, A., Navarro, J., Arrieta-Salinas, I., et al.: Improving data partition schemes in Smart Grids via clustering data streams. Expert Syst. Appl. **41**(13), 5832–5842 (2014)
14. Morrison, R.E., Bryant, C.M., Terejanu, G., et al.: Data partition methodology for validation of predictive models. Comput. Math Appl. **66**(10), 2114–2125 (2013)
15. Lloyd, S.: Least squares quantization in PCM. IEEE Trans. Inf. Theory **28**(2), 129–137 (1982)
16. Frey, B.J., Dueck, D.: Clustering by passing messages between data points. Science **315** (5814), 972 (2007)
17. Schroff, F., Kalenichenko, D., Philbin, J.: FaceNet: a unified embedding for face recognition and clustering. In: IEEE Conference on Computer Vision and Pattern Recognition, pp. 815–823. IEEE Computer Society (2015)
18. Jahirabadkar, S., Kulkarni, P.: Algorithm to determine ε-distance parameter in density based clustering. Expert Syst. Appl. **41**(6), 2939–2946 (2014)
19. Ren, Y., Liu, X., Liu, W.: DBCAMM: a novel density based clustering algorithm via using the Mahalanobis metric. Appl. Soft Comput. **12**(5), 1542–1554 (2012)
20. Guha, S., Rastogi, R., Shim, K., et al.: CURE: an efficient clustering algorithm for large databases. Inf. Syst. **26**(1), 35–58 (2001)
21. Cacciari, M., Salam, G.P., Soyez, G.: The anti-k_t jet clustering algorithm. J. High Energy Phys. **04**(4), 403–410 (2008)
22. Pal, N.R., Pal, K., Keller, J.M., et al.: A possibilistic fuzzy c-means clustering algorithm. IEEE Trans. Fuzzy Syst. **13**(4), 517–530 (2005)
23. Celebi, M.E., Kingravi, H.A., Vela, P.A.: A comparative study of efficient initialization methods for the k-means clustering algorithm. Expert Syst. Appl. **40**(1), 200–210 (2012)
24. Tran, T.N., Drab, K., Daszykowski, M.: Revised DBSCAN algorithm to cluster data with dense adjacent clusters. Chemometr. Intell. Lab. Syst. **120**(2), 92–96 (2013)

Unsupervised Maximum Margin Incomplete Multi-view Clustering

Hong Tao, Chenping Hou$^{(\boxtimes)}$, Dongyun Yi, and Jubo Zhu

National University of Defense Technology, Changsha 410073, Hunan, China
taohong.nudt@hotmail.com, hcpnudt@hotmail.com, dongyun.yi@gmail.com,
ju_bo_zhu@aliyun.com

Abstract. Discarding incomplete multi-view instances in conventional multi-view algorithms leads to a severe loss of available information. To make up for this loss, learning from multi-view incomplete data has attracted much attention. With the goal of better clustering, the Unsupervised Maximum Margin Incomplete Multi-view Clustering (UMIMC) algorithm is proposed in this paper. Different from the existing works that simply project data into a common subspace, discriminative information is incorporated into the unified representation by applying the unsupervised maximum margin criterion. Thus, the margin between different classes is enlarged in the learned subspace, leading to improvement in the clustering performance. An alternating iterative algorithm with guaranteed convergence is developed for optimization. Experimental results on several datasets verify the effectiveness of the proposed method.

Keywords: Incomplete multi-view data · Multi-view clustering
Maximum margin · Low-rank factorization

1 Introduction

Due to the increasingly growing number of multi-view data, multi-view learning has become into an important research direction in last decades [5,12,13,16,17]. Conventional multi-view learning algorithms assume that each object is fully represented by all views. However, in real-world applications, there may be some missing views for some instances [7,11,15]. For instance, in document clustering, documents are translated into different languages, but not all the documents are translated into each language; in medical diagnosis, some patients may only have the blood test, but lack the neuroimaging measurement. In traditional multi-view learning algorithms, this kind of incomplete multi-view data is generally discarded, which is a severe loss of available information [4,14,20].

To make full use of incomplete multi-view data, Li et al. proposed the Partial multi-View Clustering (PVC) algorithm [7], which is a pioneering work in the

---

This work is supported by NSF China (No. 61473302, 61503396). Chenping Hou is the corresponding author.

Z.-H. Zhou et al. (Eds.): ICAI 2018, CCIS 888, pp. 13–25, 2018.
https://doi.org/10.1007/978-981-13-2122-1_2

incomplete multi-view learning setting. Based on the Non-negative Matrix Factorization (NMF), PVC learns a latent subspace where all examples are homogeneously represented. By minimizing the differences between the consensus and the representations of each view, Shao et al. developed the Multiple Incomplete views Clustering (MIC) algorithm via weighted NMF with $L_{2,1}$ regularization [11], where incomplete examples are with smaller weights. Assuming that different views are generated from a shared subspace, Xu et al. suggested to enable the incomplete instances to be restored with the help of the complete ones by multi-view low-rank factorization [18]. Realizing that these methods neglect the geometric information of data points, Zhao et al. proposed a novel Incomplete Multi-modality Grouping (IMG) method to integrate the latent subspace generation and the compact global structure learning into a unified framework [21]. Concretely, an automatically learned graph Laplacian term is imposed on the common representation to make similar samples to be grouped together in the latent subspace. With consideration of both the inter-view and intra-view data similarities, Yin et al. regressed incomplete multi-view data to the cluster indicator matrix directly [19].

In this paper, we also focus on the incomplete multi-view clustering problem. Inspired by the Maximum Margin Criterion (MMC) [6], we want the data points belonging to different clusters to be separated with maximum margin in the shared latent subspace. More specifically, based on the low-rank assumption, a set of low-rank matrices are used to approximate the incomplete data matrices of each view. Then each approximated data matrix is factorized into the production between a view-specific basis matrix and a common representation matrix. The connections among views are considered by utilizing the same representation matrix for distinct views. To incorporate geometric information into the low-rank representation, in the latent subspace, the within-class scatter is enforced to be small while the between-class scatter to be large. By formulating both aspects into the objective, the learned common representation tends to be discriminative. The alternating iterative strategy is adopted for optimization. The effectiveness of the proposed method is validated by comparing with state-of-the-art methods on six datasets.

2   Background

Matrices and vectors are written as boldface uppercase letters and boldface lowercase letters respectively. The trace, Frobenius norm and the transpose of matrix $\mathbf{M}$ are denoted by $tr(\mathbf{M})$, $\|\mathbf{M}\|_F$ and $\mathbf{M}^T$, respectively. $\mathbf{1}$ denotes a column vector with all ones, and the identity matrix is denoted by $\mathbf{I}$.

2.1   Problem Setting

Given a data set of n instances $\mathcal{X} = \{\mathbf{x}_i\}_{i=1}^n$, its data matrix is denoted as $\mathbf{X} = [\mathbf{x}_1, \cdots, \mathbf{x}_n] \in \mathbb{R}^{d \times n}$. Suppose it has V different feature representations, i.e., $\mathbf{x}_i = [\mathbf{x}_i^{(1)}; \cdots; \mathbf{x}_i^{(V)}]$, where $\mathbf{x}_i^{(v)} \in \mathbb{R}^{d^{(v)}}$ is the i-th data point from the v-th view of $d^{(v)}$ dimension. It holds that $d = \sum_{v=1}^{V} d^{(v)}$. Let $\mathbf{X}^{(v)}$ collects the samples in the v-th view, i.e., $\mathbf{X}^{(v)} = [\mathbf{x}_1^{(v)}, \ldots, \mathbf{x}_n^{(v)}]$. In the incomplete multi-view setting, each view suffers from some missing data. That is to say, the data matrix $\mathbf{X}^{(v)}$ on each view, could have totally or partially missing columns. Thus, we define a position indicating matrix $\mathbf{\Omega}^{(v)} \in \{0,1\}^{d^{(v)} \times n}$ to record the existence of variables in $\mathbf{X}^{(v)}$ ($v = 1, \cdots, V$). Specifically, $\mathbf{\Omega}_{ij}^{(v)} = 1$ is the (i,j)-th element of $\mathbf{X}^{(v)}$ is available, and 0 otherwise.

2.2   Maximum Margin Criterion

Maximum Margin Criterion (MMC) [6] is a widely used dimension reduction method. It maximizes the average margin between classes with the low-dimensional representations, such that a pattern is close to those in the same class but far from those in different classes. Given n data points belonging to c distinct classes, MMC maximizes the following feature extraction criterion:

$$J = \frac{1}{2} \sum_{i=1}^{c} \sum_{j=1}^{c} \frac{n_i n_j}{n^2} d(\mathcal{C}_i, \mathcal{C}_j), \tag{1}$$

where n_i denotes the number of points of the i-th ($i = 1, \cdots, c$) class, and $d(\mathcal{C}_i, \mathcal{C}_j)$ is the interclass distance. By considering the scatter of classes, the interclass distance is defined as

$$d(\mathcal{C}_i, \mathcal{C}_j) = \left\| \boldsymbol{\mu}_i - \boldsymbol{\mu}_j \right\|_2^2 - tr(\mathbf{S}_i) - tr(\mathbf{S}_j), \tag{2}$$

where $\boldsymbol{\mu}_i$ and $\mathbf{S}_i$ are the mean vector and covariance matrix of the class $\mathcal{C}_i$ ($i = 1, \cdots, c$), respectively. With several mathematical operation, the following formula is obtained:

$$J = \frac{1}{n} tr(\mathbf{S}_b - \mathbf{S}_w), \tag{3}$$

where $\mathbf{S}_b$ and $\mathbf{S}_w$ are the between-class and within-class scatter matrices, respectively. They are defined as

$$\mathbf{S}_b = \sum_{i=1}^{c} n_i (\boldsymbol{\mu}_i - \boldsymbol{\mu})(\boldsymbol{\mu}_i - \boldsymbol{\mu})^T, \qquad \mathbf{S}_w = \sum_{i=1}^{c} \sum_{j=1}^{n_i} (\mathbf{v}_{ij} - \boldsymbol{\mu}_i)(\mathbf{v}_{ij} - \boldsymbol{\mu}_i)^T, \tag{4}$$

where $\boldsymbol{\mu}$ is the total mean vector, and $\mathbf{v}_{ij}$ denotes the low-dimensional representation of the j-th points in the i-th class.

3   The Proposed Method

In this section, we first introduce the formulation of the proposed method, and then present the optimization algorithm and the corresponding convergence and computational complexity analysis. .

3.1 The Formulation

Low-rank assumption is widely adopted to recover the random missing values of a matrix, but it is not applicable for missing views. Using the connection between multiple views, Xu et al. [18] adapted low-rank assumption for incomplete multi-view learning (IMVL). Specifically, they tried to find a set of low-rank matrices $\{\mathbf{Z}^{(v)}\}_{v=1}^{V}$ to approximate the incomplete multi-view data $\{\mathbf{X}^{(v)}\}_{v=1}^{V}$. Concretely, IMVL minimizes $\sum_{v=1}^{V} \left\|\mathbf{\Omega}^{(v)} \odot (\mathbf{Z}^{(v)} - \mathbf{X}^{(v)})\right\|_{F}^{2}$, where $\odot$ denotes the element-wise product between matrices. Thus, the incomplete multi-view data completion problem is formulated as follows,

$$\min_{\{\mathbf{U}^{(v)}, \mathbf{Z}^{(v)}\}_{v=1}^{V}, \mathbf{V}} \sum_{v=1}^{V} \left\|\mathbf{Z}^{(v)} - \mathbf{U}^{(v)}\mathbf{V}\right\|_{F}^{2}$$
$$\text{s.t.} \quad \mathbf{\Omega}^{(v)} \odot \mathbf{Z}^{(v)} = \mathbf{\Omega}^{(v)} \odot \mathbf{X}^{(v)}, v = 1, \cdots, V, \tag{5}$$

where $\mathbf{Z}^{(v)} \in \mathbb{R}^{d^{(v)} \times n}$ with rank r ($r < \min(d^{(v)}, n)$) is the reconstructed data matrix, $\mathbf{U}^{(v)} \in \mathbb{R}^{d^{(v)} \times r}$ is the basis matrix, and $\mathbf{V} = [\mathbf{v}_1, \cdots, \mathbf{v}_n] \in \mathbb{R}^{r \times n}$ is the common coefficient matrix shared by all views. After optimizing Problem (5), a unified representation $\mathbf{V}$ can be obtained for the incomplete multi-view dataset, and the missing data can be approximated recovered.

However, the formulation in Eq. (5) does not take any geometry information into consideration, so that the learned representation may be not satisfying to the subsequent processing. Note that our goal is to group data points into different clusters, we would like to incorporate some discriminative information into the learned unified representation. Specifically, in the latent subspace, points in the same class should be close to each other, while points in different classes should be far away from each other. Evoked by the MMC, we aim to optimize the following objective:

$$\min_{\substack{\{\mathbf{U}^{(v)}, \mathbf{Z}^{(v)}\}_{v=1}^{V}, \\ \mathbf{V}, \mathbf{F}^{T}\mathbf{F}=\mathbf{I}}} \sum_{v=1}^{V} \left\|\mathbf{Z}^{(v)} - \mathbf{U}^{(v)}\mathbf{V}\right\|_{F}^{2} + \alpha(tr(\mathbf{S}_w) - tr(\mathbf{S}_b)) + \beta \left\|\mathbf{V}\right\|_{F}^{2}$$
$$\text{s.t.} \quad \mathbf{\Omega}^{(v)} \odot \mathbf{Z}^{(v)} = \mathbf{\Omega}^{(v)} \odot \mathbf{X}^{(v)}, v = 1, \cdots, V, \tag{6}$$

where $\mathbf{F} \in \mathbb{R}^{n \times c}$ is the relaxed cluster indicator matrix, with $\mathbf{F}_{ij}$ being $\frac{1}{\sqrt{n_j}}$ if $\mathbf{x}_i$ belongs to the j-th class, and 0 otherwise. $\beta \left\|\mathbf{V}\right\|_{F}^{2}$ is a regularization term, and α and β are two non-negative trade-off parameters. $\mathbf{S}_w$ and $\mathbf{S}_b$ are calculated based on $\mathbf{V}$ and $\mathbf{F}$. It can be easily verified that

$$\mathbf{S}_w = \mathbf{VH}(\mathbf{I} - \mathbf{FF}^{T})\mathbf{HV}^{T}, \qquad \mathbf{S}_b = \mathbf{VHFF}^{T}\mathbf{HV}^{T}, \tag{7}$$

where $\mathbf{I}$ is the $n \times n$ identity matrix, and $\mathbf{H} = \mathbf{I} - \frac{1}{n}\mathbf{11}^{T}$ is the centering matrix. Denote $\mathbf{L}_b = \mathbf{HFF}^{T}\mathbf{H}$, the objective function can be rewritten as

$$\min_{\substack{\{\mathbf{U}^{(v)}, \mathbf{Z}^{(v)}\}_{v=1}^{V}, \\ \mathbf{V}, \mathbf{F}^{T}\mathbf{F}=\mathbf{I}}} \sum_{v=1}^{V} \left\|\mathbf{Z}^{(v)} - \mathbf{U}^{(v)}\mathbf{V}\right\|_{F}^{2} + \alpha tr(\mathbf{V}(\mathbf{H} - 2\mathbf{L}_b)\mathbf{V}^{T}) + \beta \left\|\mathbf{V}\right\|_{F}^{2}$$
$$\text{s.t.} \quad \mathbf{\Omega}^{(v)} \odot \mathbf{Z}^{(v)} = \mathbf{\Omega}^{(v)} \odot \mathbf{X}^{(v)}, v = 1, \cdots, V. \tag{8}$$

For presentation convenience, we refer to the proposed method as the Unsupervised Maximum Margin Incomplete Multi-view Clustering (UMIMC) algorithm.

3.2 Optimization

It can be seen that Eq. (8) has four groups of variables and is not convex. It is hard to optimize them simultaneously. The alternating minimization strategy is adopted for solving Eq. (8). Given the current iterates $\{\mathbf{U}^{(v)}\}_{v=1}^V$, $\{\mathbf{Z}^{(v)}\}_{v=1}^V$, $\mathbf{V}$ and $\mathbf{F}$, the algorithm updates these variables by minimizing Eq. (8) with respect to (w.r.t.) each one separately while fixing the others.

Optimize $\{\mathbf{U}^{(v)}\}_{v=1}^V$. The subproblem related to $\mathbf{U}^{(v)}$ is

$$\min_{\mathbf{U}^{(v)}\in\mathbb{R}^{d^{(v)}\times r}} \left\|\mathbf{Z}^{(v)} - \mathbf{U}^{(v)}\mathbf{V}\right\|_F^2, \tag{9}$$

which is a least square regression problem. The optimal solution is

$$\mathbf{U}^{(v)} = \mathbf{Z}^{(v)}\mathbf{V}^T(\mathbf{V}\mathbf{V}^T)^\dagger, \tag{10}$$

where $\mathbf{A}^\dagger$ is the Moore-Penrose pseudo-inverse of $\mathbf{A}$.

Optimize $\mathbf{V}$. Set the derivative w.r.t. $\mathbf{V}$ to zero, we have

$$\left(\sum_{v=1}^V (\mathbf{U}^{(v)})^T\mathbf{U}^{(v)} + \beta\mathbf{I}\right)\mathbf{V} + \mathbf{V}(\alpha\mathbf{H} - 2\alpha\mathbf{L}_b) = \sum_{v=1}^V (\mathbf{U}^{(v)})^T\mathbf{Z}^{(v)}, \tag{11}$$

which is a standard Sylvester equation [1]. Equation (11) has a unique solution when the eigenvalues of $\sum_{v=1}^V (\mathbf{U}^{(v)})^T\mathbf{U}^{(v)} + \beta\mathbf{I}$ and $\alpha(\mathbf{H} - 2\mathbf{L}_b)$ are disjoint.

Optimize $\{\mathbf{Z}^{(v)}\}_{v=1}^V$. The subproblem w.r.t. $\mathbf{Z}^{(v)}$ is

$$\begin{aligned} &\min_{\mathbf{Z}^{(v)}\in\mathbb{R}^{d^{(v)}\times n}} \left\|\mathbf{Z}^{(v)} - \mathbf{U}^{(v)}\mathbf{V}\right\|_F^2, \\ &\text{s.t.} \qquad \mathbf{\Omega}^{(v)} \odot \mathbf{Z}^{(v)} = \mathbf{\Omega}^{(v)} \odot \mathbf{X}^{(v)}. \end{aligned} \tag{12}$$

Thus, $\mathbf{Z}^{(v)}$ is updated by

$$\mathbf{Z}^{(v)} = \mathbf{U}^{(v)}\mathbf{V} + \mathbf{\Omega}^{(v)} \odot (\mathbf{X}^{(v)} - \mathbf{U}^{(v)}\mathbf{V}). \tag{13}$$

From Eq. (13), it can be known that the optimal $\mathbf{Z}^{(v)}$ is a duplicate of $\mathbf{X}^{(v)}$, and the missing values are filled by those of $\mathbf{U}^{(v)}\mathbf{V}$.

Optimize $\mathbf{F}$. It is equivalent to solve the following problem:

$$\max_{\mathbf{F}^T\mathbf{F}=\mathbf{I}} tr(\mathbf{F}^T\mathbf{H}\mathbf{V}^T\mathbf{V}\mathbf{H}\mathbf{F}), \tag{14}$$

which can be solved by spectral decomposition. That is to say, the optimal $\mathbf{F}$ is formed by the eigenvectors corresponding to the c largest eigenvalues of the matrix $\mathbf{H}\mathbf{V}^T\mathbf{V}\mathbf{H}$.

The pseudo code of the optimization process is summarized in Algorithm 1.

3.3   Algorithm Convergence and Computational Complexity

As shown in Sect. 3.2, each iteration of Algorithm 1 obtains the optimal solution of the corresponding subproblem. That is, the objective value of Eq. (8) is not increasing after each iteration. Thus, we have the following conclusion.

Proposition 1. *The objective function (8) is monotonically decreased with Algorithm 1.*

Proof. For convenience, denote the objective function (8) as $\mathcal{J}(\mathbf{Z}^{(v)}, \mathbf{U}^{(v)}, \mathbf{V}, \mathbf{F})$. Suppose after the k-th iteration, we have obtained $\{\mathbf{Z}_k^{(v)}, \mathbf{U}_k^{(v)}\}_{v=1}^{V}, \mathbf{V}_k$, and $\mathbf{F}_k$. According to Algorithm 1,

$$
\mathbf{U}_{k+1}^{(v)} = \underset{\mathbf{U}^{(v)} \in \mathbb{R}^{d^{(v)} \times r}}{\arg\min} \left\| \mathbf{Z}_k^{(v)} - \mathbf{U}^{(v)} \mathbf{V}_k \right\|_F^2, \tag{15}
$$

thus, we have

$$
\mathcal{J}(\{\mathbf{Z}_k^{(v)}\}_{v=1}^{V}, \{\mathbf{U}_{k+1}^{(v)}\}_{v=1}^{V}, \mathbf{V}_k, \mathbf{F}_k) \leq \mathcal{J}(\{\mathbf{Z}_k^{(v)}\}_{v=1}^{V}, \{\mathbf{U}_k^{(v)}\}_{v=1}^{V}, \mathbf{V}_k, \mathbf{F}_k). \tag{16}
$$

Then, with fixed $\{\mathbf{U}_{k+1}^{(v)}\}_{v=1}^{V}, \{\mathbf{Z}_k^{(v)}\}_{v=1}^{V}$, and $\mathbf{F}_k$, the obtained $\mathbf{V}_{k+1}$ is the unique solution of Eq. (11). It holds that

$$
\mathcal{J}(\{\mathbf{Z}_k^{(v)}\}_{v=1}^{V}, \{\mathbf{U}_{k+1}^{(v)}\}_{v=1}^{V}, \mathbf{V}_{k+1}, \mathbf{F}_k) \leq \mathcal{J}(\{\mathbf{Z}_k^{(v)}\}_{v=1}^{V}, \{\mathbf{U}_{k+1}^{(v)}\}_{v=1}^{V}, \mathbf{V}_k, \mathbf{F}_k). \tag{17}
$$

Similarly, we have

$$
\begin{aligned}
&\mathcal{J}(\{\mathbf{Z}_{k+1}^{(v)}\}_{v=1}^{V}, \{\mathbf{U}_{k+1}^{(v)}\}_{v=1}^{V}, \mathbf{V}_{k+1}, \mathbf{F}_{k+1}) \\
&\leq \mathcal{J}(\{\mathbf{Z}_{k+1}^{(v)}\}_{v=1}^{V}, \{\mathbf{U}_{k+1}^{(v)}\}_{v=1}^{V}, \mathbf{V}_{k+1}, \mathbf{F}_k) \\
&\leq \mathcal{J}(\{\mathbf{Z}_k^{(v)}\}_{v=1}^{V}, \{\mathbf{U}_{k+1}^{(v)}\}_{v=1}^{V}, \mathbf{V}_{k+1}, \mathbf{F}_k).
\end{aligned} \tag{18}
$$

Combining all the inequalities, it arrives at

$$
\mathcal{J}(\{\mathbf{Z}_{k+1}^{(v)}\}_{v=1}^{V}, \{\mathbf{U}_{k+1}^{(v)}\}_{v=1}^{V}, \mathbf{V}_{k+1}, \mathbf{F}_{k+1}) \leq \mathcal{J}(\{\mathbf{Z}_k^{(v)}\}_{v=1}^{V}, \{\mathbf{U}_k^{(v)}\}_{v=1}^{V}, \mathbf{V}_k, \mathbf{F}_k). \tag{19}
$$

That is to say, the objective function (8) is monotonically decreased with Algorithm 1.

As shown from Proposition 1, if Eq. (8) is lower-bounded, then Algorithm 1 will converge to a stationary point.

In the following, we analyze the time complexity of Algorithm 1. Algorithm 1 has four steps. The optimization of $\{\mathbf{U}^{(v)}\}_{v=1}^{V}$ involves calculating the Moore-Penrose pseudo-inverse of $\mathbf{V}\mathbf{V}^T$ and the production between matrices. There computational complexities are $\mathcal{O}(r^3)$ and $\mathcal{O}(nrd^{(v)})$, respectively. In the second step, the optimal $\mathbf{V}$ is obtained by solving a Sylvester equation, which costs $\mathcal{O}(r^3)$. Thus, the computation of $\{\mathbf{U}^{(v)}\}_{v=1}^{V}$ and $\mathbf{V}$ will not bring too much burden. Then, the solution of $\mathbf{Z}^{(v)}$ takes $\mathcal{O}(nrd^{(v)})$. Since the optimization of $\mathbf{F}$ needs to perform eigen-decomposition, its computational complexity is $\mathcal{O}(n^3)$. Note that the low rank parameter r is required to be smaller than $\min(\{d^{(v)}\}, n))$, and is usually set as a small integer (such as the number of clusters). Overall, the time complexity of each iteration of Algorithm 1 is $\mathcal{O}(n \max\{n^2, dr\})$.

Algorithm 1. UMIMC

Input: Data matrices $\{\mathbf{X}^{(v)} \in \mathbb{R}^{d^{(v)} \times n}\}_{v=1}^{V}$, the position indicating matrices $\{\mathbf{\Omega}^{(v)} \in \mathbb{R}^{d^{(v)} \times n}\}_{v=1}^{V}$, and parameters $\alpha > 0$, $\beta > 0$.
Output: $\mathbf{Z}^{(v)}, \mathbf{U}^{(v)}(\forall v \in [1, V])$, $\mathbf{V}$ and $\mathbf{F}$.
Initialization: Initialize $\mathbf{Z}_0^{(v)}(\forall v \in [1, V])$ and $\mathbf{V}_0$ randomly, set $k = 0$.
Procedure:
Repeat
1: Update $\mathbf{U}_{k+1}^{(v)}$ by Eq. (10), $\forall v \in [1, V]$
2: Update $\mathbf{V}_{k+1}$ according to Eq. (11).
3: Update $\mathbf{Z}_{k+1}^{(v)}$ by Eq. (13).
4: Update $\mathbf{F}_{k+1}$ by solving problem in Eq. (14).
5: $k = k + 1$.
Until convergence

---

4 Experiment

In this section, we first conduct experiment to verify the effectiveness of the proposed UMIMC. Then we study the impact of parameters and finally present the results about the convergence.

4.1 Data Descriptions and Experimental Settings

In this paper, six different real-world datasets are adopted to verify the effectiveness of the proposed method. They are Caltech101[1], ORL[2], Yale[3], Microsoft Research Cambridge Volume 1[4] (MSRC-v1), Handwritten digits (Digits)[5], and WebKB[6]. Their detailed descriptions are presented as follows. Note that the dimensionality of a feature type is denoted in the brackets.

- Caltech101 is an object recognition dataset consisting of 101 categories of images. We use the selected subset by [3], which has 441 images belonging to 7 classes. We refer it as **Caltech7**. SIFT (200) [8] and SURF (200) [2] features are extracted.
- **ORL** contains images of 40 distinct subjects, and each subject has 10 different images. The images were taken at different times, with varying lighting, facial expressions and facial details. We extract the SIFT (50) and wavelet texture features (32).
- The **Yale** dataset contains 165 grayscale images of 15 individuals. There are 11 images per subject, one per different facial expression or configuration. SIFT (50) and GIST (512) [10] features are extracted for experiments.

---

[1] http://www.vision.caltech.edu/Image_Datasets/Caltech101/.
[2] http://www.cad.zju.edu.cn/home/dengcai/Data/FaceData.html.
[3] http://vision.ucsd.edu/content/yale-face-database.
[4] https://www.microsoft.com/en-us/research/project/imageunderstanding/.
[5] http://archive.ics.uci.edu/ml/datasets/Multiple+Features.
[6] http://www.cs.cmu.edu/~webkb/.

- **MSRC-v1** contains 240 images in 8 classes. Following [3], the background class is discarded. We use the SIFT (200) and LBP (256) [9] features.
- **Digits** is composed of 2000 handwritten numerals from 0 to 9. Each digit corresponds to a class, and each class has 200 data points. We use 2 of 6 published features: profile correlations (216) and Fourier coefficients of the character shapes (76).
- The **WebKB** dataset consists of 1051 web pages belonging to two classes: course and non-course. It has two views: the links (2949) and the text (334).

We compare the proposed UMIMC with several state-of-the-art methods, they are PVC [7], MIC [11], IMVL [18], IMG [21] and Robust Multi-View K-Means Clustering (RMKMC) [3]. The first 4 approaches are designed in the incomplete multi-view learning setting. RMKMC integrates data's multiple representations via structured sparsity-inducing norm to make it more robust to data outliers. Since RMKMC is not designed for incomplete multi-view data, the missing data is filled with the mean of the observed samples before performing RMKMC.

For all the compared methods, the dimensionality of the latent subspace is set as the number of clusters. According to the original papers, the λ in PVC is set as 0.01, the co-regularization parameters $\{\alpha_i\}$ and the robust parameters $\{\beta_i\}$ of MIC are set to 0.01 and 0.1 for all views, and the trade-off parameters λ, γ and β in IMG are set as 0.01, 100 and 1, respectively. The parameter γ (≥ 1) in RMKMC is tuned from $\{2, 3, 5\}$. For UMIMC, α and β are set as 0.001 and 0.1, respectively.

Note that both UMIMC and IMG have two choices to do clustering: UMIMC can perform k-means on the unified representation $\mathbf{V}$ or relaxed cluster indicator matrix $\mathbf{F}$, and IMG can perform k-means on the learned representation or spectral clustering on the learned affinity graph. To make fair comparisons, we run k-means on the learned representations for both methods.

As all these datasets are originally complete, we randomly delete instances from each view to make the views incomplete. Concretely, we randomly select 10% to 50% examples and randomly remove one view from these examples. To avoid randomness, this procedure is repeated 10 times. The results are evaluated by two metrics, the normalized mutual information (NMI) and the adjusted rand index (AdjRI). Since all the compared methods need to perform k-means to get the final clustering results. We run k-means 20 times and adopt the mean as the result for each independent repetition. Finally, we report the average performance of 10 independent repetitions.

4.2 Clustering Results

Except for the average performance, we also do the Wilcoxon signed-rank test with confidence level 0.05. The comparison results are displayed in Tables 1 and 2. From the tables, we have the following observations.

As the partial example ratio (PER) increases, in terms of both NMI and AdjRI, the performance of all the compared methods is degenerated in most cases, which is consistent with the intuition. An exception is that PVC and RMKMC perform better when PER is 0.2 than that of 0.1. The possible reason is that both views of WebKB are sparse, the lack of certain points would affect the data structure largely. This could also explain why the standard deviations of IMVL and RMKMC on this dataset are large.

Each of the compared methods obtains good performance on certain datasets, but performs worse on other datasets. The possible reason is following. Apart from the subspace learning term, all the compared methods have distinct characteristics due to their different regularization terms. For example, PVC imposes non-negative constraints and ℓ_1 sparsity on the learned representation, both MIC and RMKMC use the $\ell_{2,1}$ regularization, and IMG adopts the graph Laplacian regularization. As a result, they may be good at clustering certain kinds of data, while poor at grouping the others.

As shown from the win/tie/loss counts in the last row of both tables, the proposed UMIMC can outperform the other compared methods in most cases. This is may be because that UMIMC incorporates the discriminative information into the learned representation, while the other methods do not.

4.3 Parameter Study

In this subsection, we study the impact of parameters on the proposed UMIMC. There are two non-negative parameters in UMIMC, i.e., α and β. We select values for both parameters from $\{10^{-4}, 10^{-3}, 10^{-2}, 10^{-1}, 1\}$. Without loss of generality, we train the model when the PER is 0.2. The experiments are conducted on datasets Caltech7, MSRC-v1 and Yale. Figure 1 shows the performance of UMIMC w.r.t. NMI with different parameter values.

As seen from the results, the value variation of α has larger effect on UMIMC than that of β. Although UMIMC obtains the best results for different datasets with different α values, it can achieves a relatively good performance when α is in the range of $\{10^{-4}, 10^{-3}, 10^{-2}\}$. As for β, UMIMC can get steady performance as its value varies in the candidate set.

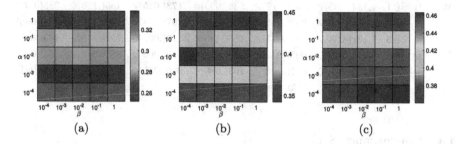

Fig. 1. Effect of parameters α and β w.r.t. NMI on three datasets: (a) Caltech7. (b) MSRC-v1. (c) Yale. Different colors means different NMI values and the color close to red indicates high scores. (Color figure online)

Table 1. Comparison results w.r.t. NMI (mean(std.)) under different PERs. Symbols '✓/◊/▼' denote that the proposed UMIMC is better/tied/worse than the corresponding method by the Wilcoxon signed-rank test with confidence level 0.05, respectively. The highest scores are in bold. The win/tie/loss counts are summarized in the last row.

Datasets	PER	PVC	MIC	IMVL	IMG	RMKMC	UMIMC
Caltech7	0.1	.236(.023)✓	.345(.015)✓	.380(.029)✓	.206(.007)✓	.423(.019)✓	.461(.012)
	0.2	.234(.022)✓	.325(.013)✓	.348(.026)✓	.355(.011)✓	.399(.019)✓	.444(.016)
	0.3	.217(.013)✓	.297(.011)✓	.329(.031)✓	.182(.007)✓	.347(.027)✓	.433(.017)
	0.4	.220(.021)✓	.280(.015)✓	.340(.031)✓	.330(.014)✓	.283(.044)✓	.425(.022)
	0.5	.234(.030)✓	.265(.020)✓	.328(.025)✓	.163(.009)✓	.277(.048)✓	.412(.015)
MSRC-v1	0.1	.459(.031)✓	.490(.023)✓	.454(.063)✓	.497(.011)✓	.466(.041)✓	.600(.019)
	0.2	.436(.026)✓	.434(.028)✓	.482(.080)◊	.482(.015)✓	.420(.023)✓	.551(.031)
	0.3	.433(.036)✓	.399(.024)✓	.459(.039)✓	.476(.013)✓	.393(.035)✓	.537(.017)
	0.4	.418(.049)✓	.379(.023)✓	.474(.053)◊	.464(.015)◊	.340(.038)✓	.513(.034)
	0.5	.416(.029)✓	.360(.029)✓	.440(.035)✓	.440(.024)✓	.331(.057)✓	.499(.025)
ORL	0.1	.567(.009)✓	.566(.010)✓	.552(.013)✓	.533(.010)✓	.555(.195)◊	.625(.010)
	0.2	.553(.011)✓	.537(.011)✓	.543(.015)✓	.507(.006)✓	.535(.189)◊	.608(.009)
	0.3	.526(.012)✓	.516(.005)✓	.537(.014)✓	.481(.009)✓	.529(.016)✓	.589(.018)
	0.4	.522(.008)✓	.489(.009)✓	.514(.023)✓	.459(.009)✓	.497(.015)✓	.568(.009)
	0.5	.498(.009)✓	.474(.006)✓	.495(.035)✓	.437(.008)✓	.465(.006)✓	.551(.013)
Yale	0.1	.370(.012)✓	.448(.012)✓	.468(.023)✓	.451(.009)✓	.424(.026)✓	.546(.015)
	0.2	.369(.017)✓	.442(.016)✓	.472(.019)✓	.452(.014)✓	.422(.026)✓	.529(.020)
	0.3	.353(.015)✓	.422(.023)✓	.461(.019)✓	.450(.015)✓	.381(.042)✓	.523(.019)
	0.4	.337(.014)✓	.391(.021)✓	.458(.029)✓	.444(.020)✓	.373(.024)✓	.505(.017)
	0.5	.329(.017)✓	.369(.023)✓	.463(.027)◊	.440(.022)✓	.350(.042)✓	.488(.023)
Digits	0.1	.607(.021)✓	.646(.010)✓	.659(.031)✓	.647(.005)✓	.639(.017)✓	.691(.008)
	0.2	.537(.022)✓	.588(.005)✓	.633(.053)✓	.619(.006)✓	.618(.014)✓	.680(.008)
	0.3	.524(.026)✓	.552(.014)✓	.600(.043)✓	.605(.008)✓	.471(.014)✓	.641(.016)
	0.4	.496(.020)✓	.518(.016)✓	.587(.047)✓	.593(.004)✓	.419(.014)✓	.616(.013)
	0.5	.473(.029)✓	.491(.017)✓	.579(.089)◊	.572(.007)✓	.363(.013)✓	.607(.016)
WebKB	0.1	.134(.035)✓	.052(.017)✓	.238(.095)◊	.173(.015)✓	.135(.190)◊	.288(.092)
	0.2	.143(.010)✓	.032(.011)✓	.157(.122)◊	.167(.034)✓	.186(.150)◊	.261(.068)
	0.3	.131(.024)✓	.011(.013)✓	.155(.133)◊	.117(.038)◊	.150(.155)◊	.183(.083)
	0.4	.112(.018)◊	.005(.004)✓	.114(.093)◊	.119(.065)◊	.029(.048)✓	.136(.043)
	0.5	.108(.026)◊	.001(.002)✓	.103(.086)◊	.086(.065)◊	.034(.049)✓	.103(.041)
Win/tie/loss		28/2/0	30/0/0	21/9/0	26/4/0	25/5/0	-

4.4 Convergence Study

In Sect. 3.3, we have shown that Algorithm 1 will monotonically decrease the objective value of Eq. (8). If Eq. (8) is lower-bounded, then Algorithm 1 will converge. In this subsection, we conduct experiments to see the convergence behavior of Algorithm 1. Concretely, experiments are performed on datasets Catech7,

Table 2. Comparison results w.r.t. AdjRI (mean(std.)) with different PERs (see the title of Table 1 for more information).

Datasets	PER	PVC	MIC	IMVL	IMG	RMKMC	UMIMC
Caltech7	0.1	.127(.016)✓	.210(.012)✓	.285(.030)✓	.107(.004)✓	.293(.013)✓	.353(.008)
	0.2	.126(.020)✓	.194(.011)✓	.247(.033)✓	.227(.012)✓	.277(.044)✓	.342(.019)
	0.3	.112(.011)✓	.167(.010)✓	.237(.032)✓	.087(.008)✓	.201(.052)✓	.339(.026)
	0.4	.112(.015)✓	.141(.011)✓	.249(.034)✓	.212(.013)✓	.138(.064)✓	.323(.021)
	0.5	.122(.024)✓	.121(.019)✓	.239(.028)✓	.066(.010)✓	.138(.070)✓	.316(.016)
MSRC-v1	0.1	.312(.042)✓	.355(.028)✓	.323(.070)✓	.372(.015)✓	.346(.050)✓	.484(.027)
	0.2	.282(.034)✓	.289(.032)✓	.359(.080)◊	.354(.017)✓	.268(.034)✓	.434(.039)
	0.3	.300(.041)✓	.243(.022)✓	.335(.038)✓	.350(.012)✓	.218(.039)✓	.415(.028)
	0.4	.290(.057)✓	.204(.024)✓	.352(.059)◊	.326(.024)◊	.157(.038)✓	.385(.044)
	0.5	.285(.036)✓	.183(.026)✓	.311(.037)✓	.309(.024)✓	.134(.037)✓	.374(.036)
ORL	0.1	.149(.010)✓	.139(.011)✓	.139(.013)✓	.107(.010)✓	.182(.065)✓	.219(.013)
	0.2	.130(.011)✓	.102(.008)✓	.132(.015)✓	.080(.005)✓	.150(.057)✓	.200(.010)
	0.3	.097(.013)✓	.076(.005)✓	.130(.012)✓	.057(.006)✓	.074(.013)✓	.183(.020)
	0.4	.087(.011)✓	.055(.006)✓	.112(.017)✓	.043(.006)✓	.051(.009)✓	.157(.008)
	0.5	.062(.009)✓	.042(.004)✓	.097(.021)✓	.030(.004)✓	.031(.003)✓	.140(.014)
Yale	0.1	.092(.011)✓	.162(.010)✓	.190(.024)✓	.169(.010)✓	.125(.026)✓	.275(.018)
	0.2	.090(.014)✓	.152(.014)✓	.193(.021)✓	.169(.015)✓	.116(.028)✓	.257(.024)
	0.3	.072(.014)✓	.134(.022)✓	.183(.021)✓	.168(.017)✓	.073(.025)✓	.247(.024)
	0.4	.057(.010)✓	.102(.017)✓	.182(.029)✓	.163(.019)✓	.075(.021)✓	.230(.020)
	0.5	.045(.013)✓	.083(.017)✓	.184(.031)◊	.162(.022)✓	.061(.030)✓	.212(.026)
Digits	0.1	.470(.025)✓	.523(.014)✓	.576(.039)✓	.533(.008)✓	.516(.027)✓	.614(.012)
	0.2	.402(.022)✓	.441(.009)✓	.548(.060)◊	.504(.009)✓	.491(.024)✓	.600(.013)
	0.3	.389(.030)✓	.383(.017)✓	.515(.048)✓	.481(.014)✓	.228(.025)✓	.558(.024)
	0.4	.347(.019)✓	.334(.021)✓	.496(.056)✓	.464(.008)✓	.149(.020)✓	.531(.020)
	0.5	.320(.031)✓	.292(.022)✓	.482(.110)◊	.438(.009)✓	.112(.009)✓	.517(.020)
WebKB	0.1	.179(.078)✓	.123(.050)✓	.364(.123)◊	.329(.019)✓	.178(.254)✓	.401(.111)
	0.2	.199(.021)◊	.098(.028)✓	.239(.187)◊	.319(.042)✓	.260(.219)◊	.364(.087)
	0.3	.201(.029)✓	.055(.036)✓	.205(.225)◊	.242(.069)✓	.239(.219)◊	.274(.110)
	0.4	.163(.027)◊	.033(.023)✓	.155(.181)◊	.231(.122)◊	-.013(.023)✓	.222(.059)
	0.5	.162(.045)◊	.000(.014)✓	.143(.121)◊	.176(.116)◊	-.021(.041)✓	.156(.047)
Win/tie/loss		27/3/0	30/0/0	20/10/0	27/3/0	28/2/0	-

MSRC-v1 and Yale when the PER is 0.2. We plot the curves of objective values of UMIMC in Fig. 2. As we can see, on these three datasets, the objective function of UMIMC converges to a fixed value after several iterations. On Caltech7 and Yale, the algorithm converges within 30 iterations. For MSRC-v1, no more than 40 iterations are needed for convergence.

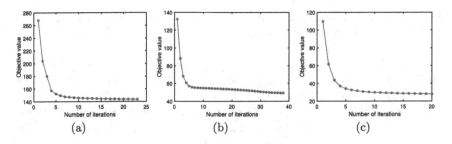

Fig. 2. Curves of objective function of UMIMC on three datasets: (a) Caltech7. (b) MSRC-v1. (c) Yale.

5 Conclusion

In this paper, we propose a method to deal with the incomplete multi-view data with consideration of the discriminative information. By combining the low-rank matrix factorization approximation term and the unsupervised maximum margin term, we propose the Unsupervised Maximum Margin Incomplete Multi-view Clustering (UMIMC) algorithm. UMIMC is not only able to learn a unified representation for the multi-view incomplete data, but also locate different classes with maximum margin in the latent space. As a result, better clustering performance can be achieved with the representation learned by UMIMC. Experimental results on six datasets validate the effectiveness of UMIMC.

References

1. Bartels, R.H., Stewart, G.W.: Solution of the matrix equation AX + XB = C. Commun. ACM **15**(9), 820–826 (1972)
2. Bay, H., Tuytelaars, T., Van Gool, L.: SURF: speeded up robust features. In: Leonardis, A., Bischof, H., Pinz, A. (eds.) ECCV 2006. LNCS, vol. 3951, pp. 404–417. Springer, Heidelberg (2006). https://doi.org/10.1007/11744023_32
3. Cai, X., Nie, F., Huang, H.: Multi-view K-means clustering on big data. In: IJCAI, pp. 2598–2604 (2013)
4. Hou, C., Nie, F., Tao, H., Yi, D.: Multi-view unsupervised feature selection with adaptive similarity and view weight. IEEE Trans. Knowl. Data Eng. **29**(9), 1998–2011 (2017)
5. Hou, C., Zhang, C., Wu, Y., Nie, F.: Multiple view semi-supervised dimensionality reduction. Pattern Recogn. **43**(3), 720–730 (2010)
6. Li, H., Jiang, T., Zhang, K.: Efficient and robust feature extraction by maximum margin criterion. IEEE Trans. Neural Netw. **17**(1), 157–165 (2006)
7. Li, S.Y., Jiang, Y., Zhou, Z.H.: Partial multi-view clustering. In: AAAI, pp. 1968–1974 (2014)
8. Lowe, D.G.: Distinctive image features from scale-invariant keypoints. Int. J. Comput. Vis. **60**(2), 91–110 (2004)
9. Ojala, T., Pietikäinen, M., Mäenpää, T.: Multiresolution gray-scale and rotation invariant texture classification with local binary patterns. IEEE Trans. Pattern Anal. Mach. Intell. **24**(7), 971–987 (2002)

10. Oliva, A., Torralba, A.: Modeling the shape of the scene: a holistic representation of the spatial envelope. Int. J. Comput. Vis. **42**(3), 145–175 (2001)
11. Shao, W., He, L., Yu, P.S.: Multiple incomplete views clustering via weighted non-negative matrix factorization with $L_{2,1}$ regularization. In: Appice, A., Rodrigues, P.P., Santos Costa, V., Soares, C., Gama, J., Jorge, A. (eds.) ECML PKDD 2015. LNCS (LNAI), vol. 9284, pp. 318–334. Springer, Cham (2015). https://doi.org/10. 1007/978-3-319-23528-8_20
12. Sun, S.: A survey of multi-view machine learning. Neural Comput. Appl. **23**(7), 2031–2038 (2013)
13. Tao, H., Hou, C., Nie, F., Zhu, J., Yi, D.: Scalable multi-view semi-supervised classification via adaptive regression. IEEE Trans. Image Process. **26**(9), 4283–4296 (2017)
14. Tao, H., Hou, C., Zhu, J., Yi, D.: Multi-view clustering with adaptively learned graph. In: ACML, pp. 113–128 (2017)
15. Trivedi, A., Rai, P., Daumé III, H., DuVall, S.L.: Multiview clustering with incomplete views. In: NIPS Workshop (2010)
16. Xu, C., Tao, D., Xu, C.: Large-margin multi-view information bottleneck. IEEE Trans. Pattern Anal. Mach. Intell. **36**(8), 1559–72 (2014)
17. Xu, C., Tao, D., Xu, C.: A survey on multi-view learning. arXiv preprint:1304.5634 (2013)
18. Xu, C., Tao, D., Xu, C.: Multi-view learning with incomplete views. IEEE Trans. Image Process. **24**(12), 5812–5825 (2015)
19. Yin, Q., Wu, S., Wang, L.: Unified subspace learning for incomplete and unlabeled multi-view data. Pattern Recogn. **67**(67), 313–327 (2017)
20. Zhang, L., Zhao, Y., Zhu, Z., Shen, D., Ji, S.: Multi-view missing data completion. IEEE Trans. Knowl. Data Eng. **PP**(99), 1 (2018)
21. Zhao, H., Liu, H., Fu, Y.: Incomplete multi-modal visual data grouping. In: IJCAI, pp. 2392–2398 (2016)

Multi-view K-Means Clustering with Bregman Divergences

Yan Wu[1,2(✉)], Liang Du[1,2(✉)], and Honghong Cheng[1,2]

[1] The School of Computer and Information Technology, Shanxi University,
Taiyuan 030006, Shanxi Province, China
18335103184@163.com, csliangdu@gmail.com
[2] Institute of Big Data Science and Industry, Shanxi University, Taiyuan 030006,
Shanxi Province, China

Abstract. Multi-view clustering has become an important task in machine learning. How to make full use of the similarities and differences among multiple views to generate clusters is a crucial issue. However, the existing multi-view clustering methods rarely consider the redundancy of the multiple views. In this paper, we propose a novel multi-view clustering method with Bregman divergences (MVBDC), where the clustering result is achieved by minimizing the Bregman divergences between clustering results obtained by weighted multiple views and the item that controls redundancy of multiple views. The experimental results on nine data sets demonstrate that our algorithm has a good clustering performance.

Keywords: Multi-view clustering · Bregman divergences · K-means

1 Introduction

Due to the fact that many of the data in real life are portrayed from multiple perspectives, for example, some news are reported from various sources, exploring useful knowledge from these data is a major topic of data mining [27]. Most of the multi-view data are unlabeled data, mining the useful information from the huge amounts of unlabeled data becomes an inevitable challenge [16],[25]. The k-means algorithm was proposed to extract knowledge from unlabeled data and has been easily implemented at large scale [8]. However, a drawback of the traditional k-means algorithm is that it can only be applied to single-view data. Therefore, in this paper, we propose an algorithm that can extend the k-means to the scene of multiple views.

Many multi-view clustering methods have been proposed to overcome the restriction that conventional clustering algorithms can only cluster single view

Supported by Chinese National Natural Science Young Foundation: Robust Clustering Models and Algorithms for Multi-source Big Data, Number: 61502289, 2016.01-2018.12.

[3,5,6,17,20]. Existing multi-view clustering methods can be roughly divided into two branches according to the fashion that multiple views are used, namely, distributed methods and centralized methods. The former methods cluster individual views separately and integrate multiple clustering results and the latter methods take full advantage of the hidden relationship among multiple views to find the cluster during the process of clustering. Then a problem arises, *i.e.* how to make full use of information to produce a division taking into account the differences and relevance among the multiple views. However, most of the existing methods give each view the same weight, which may result in poor clustering result due to unimportant or noisy views. Kernel-based methods [28] and NMF-based methods [10,14,21,29,30] are proposed to solve these problems.

The Kernel-based methods including kernel k-means [9] and spectral clustering [22] are widely used and presented with good performance because they can capture the non-linear structure of data sets, however, these methods usually lead to high computational complexity because of eigen decomposition [19] or complex selection of kernels and parameters. By contrast, NMF-based methods can avoid these problems. Nonnegative matrix factorization(NMF) [12] is a technique to find two non-negative matrices whose product can well approximate the original matrix [4]. Cai [5] proposed a multi-view clustering method based on k-means, this k-means-like method [2,7,11,23,24,26] uses $L_{21}-$norm [13,15] to push clustering solution of each view towards a common consensus within a short time. However, almost none of these technologies have considered the redundancy among the views or have provided a generic paradigm for multi-view clustering.

In this work, we add the redundancy constraints on multiple views and use a new consensus criterion – Bregman divergences to solve the above problem [18]. Unlike RMKMC [5], this article uses a much more general criterion – Bregman divergences to measure the distortion among the clustering result and multiple views. Inspired by [18], we extend the classical k-means which use Euclidean distance into a multi-view scene using Bregman divergences [1]. Besides, we balance the importance of each view through maintaining a set of learnable weights for views so that each view can have the optimal complementarity to each other. Our framework of multi-view clustering is weighted and the weights that balance the importance of each view are learned automatically. In addition, an alternating iterative algorithm is provided to find the optimal clustering results.

2   Multi-view Clustering via Bregman Divergences

In this section, we systematically present a novel multi-view clustering method using Bregman divergences.

2.1   The Construction of Objective Function

In general, we assume the giving data set has n_v different views, the size of each view is $\mathbb{R}^{n \times d_v}(v = 1, 2, \cdots, n_v)$, d_v is the number of attributes in v-th view, where each row of multi-view data represents an object described by a series of attributes. Different view data in the real world are often described by different number of attributes, but the number of the object is the same for diverse views. The goal of multi-view clustering is to integrate different views and form clustering results with the integrated data. In order to achieve the target, we need to solve the challenging problem: How to uncover the relationship among various views and integrate the multi-views reasonably. More formally, we calculate the similarity between each pair of samples in each view and describe it as $\mathbb{A}^{n \times n}$. Then, separate views can be integrated together to perform the clustering algorithm. In this article, we introduce Bregman divergences to measure the information loss between single view and clustering result. In the first item, we add the weights to decrease multi-view information loss caused by clustering. In the second item, we add the diversity measurement of the information between each pair of views to reduce the redundancy, that is, to increase the difference among various data sources used in clustering. The objective function is shown in $Eq.\,(1)$.

$$\min_{U,V,(w^{(v)})} \sum_{v=1}^{n_v} w^{(v)} D\phi(UV, A^{(v)}) + \gamma w^{\mathrm{T}} H w;$$

$$\text{s.t.} \ \sum_{v=1}^{n_v} w^{(v)} = 1, \sum_{k=1}^{K} U_{ik} = 1, U_{ik} \in \{0, 1\} \tag{1}$$

where:

w – the weight vector of multi-view matrix, and $w \in \mathbb{R}_+^{n_v \times 1}$ is also described as $w = (w^{(1)}, w^{(2)}, ..., w^{(n_v)})^T$;

$D\phi(A, B)$ – the Bregman divergences that measure the information loss between A and B. U– the cluster representation, the U satisfies the $1 - of - k$ coding scheme. U is described in row vectors as $U = \{U_1, U_2, ..., U_n\}$, each row vector is $U_i^{1 \times K} (i = 1, 2, \cdots, n)$, K is the number of cluster centers;

V – V and U together constitute a clustering result matrix, $V \in \mathbb{R}_+^{K \times n}$ is described in row vectors as $V = \{V_1, V_2, ..., V_K\}$, each row vector can be described as $V_k^{1 \times n} (k = 1, 2, \cdots, K)$, n is the number of samples;

$A^{(v)}$ – the v-th view matrix which is obtained by computing the Euclidean distance between each pair of samples, there is n_v view matrices $\{A^{(1)}, A^{(2)}, ..., A^{(n_v)}\}$, and $A^{(v)} \in \mathbb{R}_+^{n \times n} (v = 1, 2, \cdots, n_v)$;

H – the independence measurement between each pair of views, $H \in \mathbb{S}_+^{n_v}$, we use $HSIC$ as the independent criterion, we shall investigate the solution of H and the convexity in Subsect. 2.3;

γ – the trade-off parameter to control the proportion of the independence of multiple views in the objective function.

In the next section, we will show the derivation of algorithm in detail.

2.2   The Solution of Objective Function

In this section, we will discuss the solution of the objective function.

The Solve of U. We solve the cluster index matrix U by decoupling the cluster data into $U_i = U_1, U_2, \cdots, U_n \in \mathbb{R}^{1 \times K}$, that is, each sample is assigned into cluster independently. We solve U by:

$$U_i^* = e_k^{\mathrm{T}}, \qquad k = \underset{j}{argmin} \sum_{v=1}^{n_v} \|A_i^{(v)} - C_j\|_2, (j = 1, 2, \cdots, K) \tag{2}$$

where the C_j is the cluster center vector, there are K cluster centers and $C_j \in \mathbb{R}^{1 \times n}$, we can send each sample to the nearest cluster and obtain the optimist cluster division of current multi-view aggregated matrix.

The Solve of V. Given matrix U, we use the objective function to derive the matrix V. Our object is to find a matrix V to minimize:

$$J_1 = \min_{V} \sum_{v=1}^{n_v} w^{(v)} D\phi(UV, A^{(v)}) \tag{3}$$

Due to the target matrix is structural, we can decouple the matrix into row vectors i.e. $U_i(i = 1, 2, \cdots, n) \in \mathbb{R}^{1 \times k}$, and $A_i(i = 1, 2, \cdots, n) \in \mathbb{R}^{1 \times n}$. First, we solve V by dividing $Eq.\,(3)$ into $Eq.\,(4)$:

$$J_1 = \min_{V_i} \sum_{v=1}^{n_v} w^{(v)} \sum_{i=1}^{n} D\phi(U_i V_i, A_i^{(v)}) \tag{4}$$

And then, we divide these row vectors into elements, the objective function can be simplified greatly in this way. Note that, the k is the position where matrix $U_{ik} = 1$, i.e. the solution of $Eq.\,(2)$ for $i = (1, 2, \cdots, n)$.

$$J_1 = \min_{V_{kj}} \sum_{v=1}^{n_v} w^{(v)} \sum_{i=1}^{n} \sum_{j=1}^{n} D\phi(U_{ik} V_{kj}, A_{ij}^{(v)}) \tag{5}$$

And $D\phi$ denotes any separable Bregman divergence and

$$D\phi(x, y) \triangleq \phi(x) + \phi(y) - \nabla\phi(y)(x - y) \tag{6}$$

So, the objective function $Eq.\,(5)$ can be described as

$$J_1 = \min_{V_{kj}} \sum_{v=1}^{n_v} w^{(v)} \sum_{i=1}^{n} \sum_{j=1}^{n} [\phi(V_{kj}) - \phi(A_{ij}^{(v)}) - \nabla\phi(A_{ij}^{(v)})(V_{kj} - A_{ij}^{(v)})] \tag{7}$$

where $\phi : S \subseteq \mathbb{R} \to \mathbb{R}$ is a strictly convex function, taking derivative of J_1 with respect to V_{kj}:

$$\frac{\partial J_1}{\partial V_{kj}} = \sum_{v=1}^{n_v} w^{(v)} \sum_{i=1}^{n} \sum_{j=1}^{n} (\nabla\phi(V_{kj}) - \nabla\phi(A_{ij}^{(v)})) \tag{8}$$

And let

$$\frac{\partial J_1}{\partial V_{kj}} = 0$$

Then we will get

$$\nabla\phi(V_{kj}) = \sum_{\substack{argU_{ik}\neq 0 \\ i}} \sum_{k=1}^{n_v} w^{(v)} \nabla\phi(A_{ij}^{(v)}) \tag{9}$$

The solution of $Eq.(9)$ is shown in Table 1.

Table 1. Bregman divergences and corresponding optimal V

	Distance	$\mathcal{D}(x)$	$\phi(x)$	$\nabla\phi(x)$	V_{kj}
1	Euclidean distance	$\mathbb{R}$	$\frac{1}{2}x^2$	x	$\sum_k \sum_{v=1}^{V_n} w^{(v)} A_{ij}^{(v)}$
2	Exponential distance	$\mathbb{R}$	e^x	e^x	$\sum_k log(\sum_{v=1}^{n_v} w^{(v)} e^{A_{ij}^{(v)}})$
3	Kullback-Leibler divergence	$\mathbb{R}_{++}$	$xlogx - x$	$\sum_{\substack{argU_{ig}\neq 0 \\ i}} logx$	$\sum_k exp(\sum_{v=1}^{n_v} w^{(v)} log(A_{ij}^{(v)}))$
4	Itakura-Saito distance	$\mathbb{R}_{++}$	$-logx$	$\frac{1}{x}$	$\sum_k (1/\sum_{v=1}^{n_v} w^{(v)} \frac{1}{A_{ij}^{(v)}})$

The Solve of V. In order to describe the difference among the information of multiple views, we use the Hilbert-Schmidt Independence Criterion(HSIC) as an independence criterion among multiple views. For example, the calculation formula of $HSIC$ between view 1 and view 2 is shown in $Eq.9$.

$$HSIC(A^{(v1)}, A^{(v2)}) = \frac{tr(MHLH)}{n^2} \tag{10}$$

where $M, L, H \in \mathbb{R}^{n\times n}$, $M = A^{(v1)} \times (A^{(v1)})$, $L = A^{v2} \times (A^{v2})'$, and $H = (I - \mathbb{1} \times \mathbb{1}^T/n)$, I is identity matrix, and $I \in \mathbb{R}_+^{n\times n}$, $\mathbb{1}$ is a vector values 1, and $\mathbb{1} \in \mathbb{R}_{++}^{n\times 1}$, $tr(\cdot)$ is the trace of matrix. It is worth mentioning that the independence criterion matrix H is positive semi-definite matrix, that is $S \in \mathbb{S}_+^p$. So the objective function about w is convex.

With matrix H, we transfer information learned from one view to the other and use it to further refine the relationships in the data, and generate an optimal clustering result finally.

2.3   Algorithm of Multi-view Clustering with Bregman Divergences

In this section, we will mainly introduce the algorithm of multi-view clustering. The objective function $Eq.(1)$ is not convex with respect in w, U and V at the same time, so we use the method of block coordinate descent algorithm [18] to tackle the problem. That is, keeping one variable fixed, then the optimization over the other is a convex problem with a unique feasible solution. In this way, we can guarantee that the objective function is monotonically descending and can converge to a stationary point.

First, when we fix w, we can get the weighted aggregation matrix easily, then the objection problem becomes a linear function we can rewrite the optimization function as $min_{U,V} D\phi(UV, A^{(v)})$. And the optimal of U can be obtained by cluster process, further, we can get the optimal V using $Eq.(9)$. Then, when the UV is fixed, $i.e. D\phi(UV, A^{(v)})$ is available, the optimal problem about w is a quadratic programming problem.

In the algorithm of multi-view clustering, we solve the objective function through iteration. We can get the solution of U^t and V^t from the weight aggregation multi-view matrix, which the weight w^t is obtained from the last iteration. And then, we use the $U^t V^t$ to find the solution of w^t by quadratic programming. In this way, we update the variables w, U, V iteratively until the cluster center or the indicator no longer change.

Algorithm 1. Multi-view clustering algorithm via bregman divergences

---

input :

      multi-view matrices $\{M_v\}_{v=1}^{n_v}$

output:

      the cluster representation matrix U^t

      the learned weight w

1   initialize the model matrix parameter $w^0 = [\frac{1}{n_v}, \frac{1}{n_v}, \cdots, \frac{1}{n_v}]$; standardized

     modal distance matrices $\{A^{(v)}\}_{v=1}^{n_v}$; initialize the aggregation modal matrix

     $A^0 = \sum_{v=1}^{n_v} \frac{1}{n_v} A^{(v)}$; $t=0$; initialize the cluster center C^0 at random; initialize the

     threshold $s=0.0001$;

2   compute H by $Eq.(0)$;

3   **repeat**

4      $t = t + 1$;

5   solve the cluster indicator matrix U^t by $Eq.(2)$;

6   solve the matrix of centroid V^t by $Eq.(9)$;

7   compute the divergence $D\phi(U^t V^t, A^{(v)^t})$ by $Eq.(6)$;

8   solve the weight vector w^t by

$$min_{w^t} \sum_{v=1}^{n_v} w^{(v)^t} D\phi(U^t V^t, A^{(v)^t}) + \gamma w^{T^t} H w^t$$

$$s.t. \sum_{v=1}^{n_v} (w^{(v)})^t = 1, 0 \le (w^{(v)})^t \le 1 (\forall v = 1, 2, \cdots, n_v)$$

9 **until** $\|U^t - U^{t-1}\|_F < s$;

2.4   Algorithm Complexity Analysis

We present here the complexity analysis of the above algorithm, in the initialization phase, the cost for A is $O(n \times d \times n_v)$, where we assume d is the dimension of view matrices, in fact each view matrix has a different dimension, and there are n_v views, noting that $n_v \ll n$. In each iteration, the worst computation cost for U by $Eq.\,(2)$ is $O(n \times n \times K)$, because we send n samples into K clusters and each sample has n dimensions. The computation cost for V by $Eq.\,(9)$ is $O(n \times n)$ for the reason that the computing of matrix V is an operation of sum, the number of elements participating in basic operations is $n \times n$ in the worst situation. And the worst computation cost for C by $Eq.(11)$ is $O(n \times n)$, similar to V, there are $n \times n$ elements participate in basic operations. Besides, the computation cost for $D\phi(UV, A^{(v)})$ by $Eq.\,(6)$ is $O(n^2)$. In all, the overall time complexity of the algorithm is $O(n \times n_v \times d + t \times K \times n^2)$.

3   Experiment

In this section, we will conduct a set of experiments to evaluate the performance of the proposed multi-view clustering algorithm. The experiment results are obtained from nine broadly used multi-view data sets which were used in [16].

3.1   Data Set Descriptions

We report the statistic of the nine data sets in Table 2.

Table 2. Description of nine public data sets

	Data sets (type of data)	Num. of objects	Num. of features	Views	Clusters
1	3sources (news)	169	{3560, 3631, 3068}	3	6
2	BBC (news)	685	{4659,4633, 4665, 4684}	4	5
3	BBCSport (news)	544	{3183, 3203}	2	5
4	WikipediaArticles (articles)	693	{128, 10}	2	10
5	Digits (image)	2000	{76, 216, 64, 240, 47, 6}	6	10
6	WebKB-texas (web page)	187	{187, 187, 187, 1703}	4	5
7	WebKB-washington (web page)	230	{230, 230, 230, 1703}	4	5
8	Cornell (web page)	195	{195, 195, 195, 1703}	4	5
9	20newsgroups (news)	500	{2000, 2000, 2000}	3	5

3.2   Experimental Setup

In this subsection, we will show the experimental setup of our proposed method and the contrast method. In order to verify the rationality of the weights added to views, we first apply every single view in our algorithm framework on the above data sets. Then, we remove the study of the weight from our algorithm and assign each view the same weight, in this way we obtain a simple version of our algorithm (SMVBDC). Next, we compare our proposed method with the robust multi-view K-means clustering(RMKMC) [5] to further investigate the effectiveness of the proposed multi-view clustering method. As we all know, clustering results are susceptible to initialization, so, following the similar experimental setup of other multi-view clustering methods, we repeat the single-view counterparts and the compared method 50 times with random initialization and show the average result to exclude the effect of initialization. For RMKMC, we strictly abide by the authors' suggestion in [5] to assign values of parameters γ, in detail, we search the $\log_{10}\gamma$ ranging from 0.1 to 2 with incremental step 0.2 to get the best parameter γ^*. For our method, we show the result of distances 1, 3 and 4 of Bregman divergences $i.e.$ the Euclidean Distance, the Kullback-Leibler Divergence and the Itakura-Saito Distance in the following tables. By searching the value of p in the range of 0.1 to 2 with step 0.2 and γ in $[10^{-5}, \cdots, 10^7]$, and we choose the best results to report (Tables 3, 4, 5, 6, 7).

3.3 Results and Discussion

In these results tables, we bold the results of the best clustering accuracy, NMI and Purity respectively. As shown in the following tables, MVBDC usually provides the best quality improvement w.r.t. clustering. First, on each data set, our experimental results are superior to the clustering results of a single view. This illustrates the advantages of using the learned weights to fuse information from multiple views. Furthermore, it outperforms equally-weighted version of our framework in most instances, which means the weight we search is better. This can further illustrate that it is useful to have redundancy constraints on multiple views in our experiments. In contrast to the competing approach, we achieve a clear cluster accuracy enhancement up to 0.4 for multi-view clustering. Besides, in the last three rows of each table, we used three different distances to characterize clustering losses. From those results, we can see that the first distance, which is the Euclidean distance, often gives the best clustering results (Tables 8, 9, 10, 11).

Table 3. 3sources data set

Method	ACC	NMI	Purity
View1 BBC	0.5551 ± 0.0619	0.5001 ± 0.0431	0.6984 ± 0.0356
View2 Reuters	0.5463 ± 0.0632	0.5110 ± 0.0447	0.6986 ± 0.0349
View3 The Guardian	0.5449 ± 0.0717	0.4775 ± 0.0522	0.6724 ± 0.0339
SMVBDC distance1	0.5893 ± 0.0484	0.5446 ± 0.0610	0.7041 ± 0.0403
SMVBDC distance3	0.5771 ± 0.0711	0.5439 ± 0.0571	0.7142 ± 0.0475
SMVBDC distance4	0.6158 ± 0.0693	0.5805 ± 0.0473	0.7396 ± 0.0171
RMKMC	0.2331 ± 0.0579	0.0489 ± 0.0197	0.3485 ± 0.0865
Our method distance1	$\mathbf{0.6432 \pm 0.0537}$	$\mathbf{0.6219 \pm 0.0839}$	$\mathbf{0.7822 \pm 0.0664}$
Our method distance3	0.6426 ± 0.0548	0.5686 ± 0.0508	0.7266 ± 0.0392
Our method distance4	0.6266 ± 0.0694	0.5652 ± 0.0443	0.7379 ± 0.0362

Table 4. BBC data set

Method	ACC	NMI	Purity
View1	0.5324 ± 0.0677	0.3761 ± 0.0497	0.6101 ± 0.5324
View2	0.5331 ± 0.0577	0.3672 ± 0.0593	0.6096 ± 0.0509
View3	0.5123 ± 0.0424	0.3507 ± 0.0449	0.6024 ± 0.0423
View4	0.5073 ± 0.0446	0.3439 ± 0.0486	0.5832 ± 0.0424
SMVBDC diatance1	0.5830 ± 0.0740	0.4979 ± 0.0714	0.6516 ± 0.0579
SMVBDC distance3	0.5836 ± 0.0286	0.4200 ± 0.0918	0.6218 ± 0.0733
SMVBDC distance4	0.6014 ± 0.0364	0.4289 ± 0.0533	0.6113 ± 0.0336
RMKMC	0.2286 ± 0.0561	0.0101 ± 0.0032	0.3129 ± 0.0747
Our method distance1	0.6266 ± 0.0694	**0.5652 ± 0.0443**	**0.7379 ± 0.0362**
Our method distance3	**0.6337 ± 0.0713**	0.5200 ± 0.0757	0.6825 ± 0.0637
Our method distance4	0.5658 ± 0.0251	0.4329 ± 0.0448	0.6378 ± 0.0554

Table 5. BBCSport data set

Method	ACC	NMI	Purity
View1	0.5561 ± 0.0796	0.3524 ± 0.0711	0.6115 ± 0.0597
View2	0.5609 ± 0.0614	0.3629 ± 0.0590	0.6278 ± 0.0496
SMVBDC diatance1	0.6746 ± 0.0824	0.4817 ± 0.0770	0.7018 ± 0.0714
SMVBDC distance3	0.6595 ± 0.0391	0.5025 ± 0.0918	0.7209 ± 0.0393
SMVBDC distance4	0.6569 ± 0.0884	0.5062 ± 0.0616	0.7227 ± 0.0459
RMKMC	0.3963 ± 0.1208	0.0584 ± 0.0751	0.4565 ± 0.1192
Our method distance1	**0.7202 ± 0.0695**	**0.5594 ± 0.0392**	**0.7597 ± 0.0357**
Our method distance3	0.7007 ± 0.0582	0.5026 ± 0.0651	0.7167 ± 0.0536
Our method distance4	0.6024 ± 0.0385	0.4128 ± 0.0598	0.6767 ± 0.0416

Table 6. WikipediaArticles data set

Method	ACC	NMI	Purity
View1	0.1890 ± 0.0079	0.0670 ± 0.0037	0.2093 ± 0.0065
View2	0.5903 ± 0.0236	0.5608 ± 0.0086	0.6350 ± 0.0123
SMVBDC diatance1	0.5829 ± 0.0393	0.5448 ± 0.0066	0.6230 ± 0.0089
SMVBDC distance3	0.5529 ± 0.0186	0.5469 ± 0.0091	0.6222 ± 0.0082
SMVBDC distance4	0.5766 ± 0.0251	0.5478 ± 0.0080	0.6256 ± 0.0087
RMKMC	0.3918 ± 0.1015	0.3373 ± 0.0882	0.4410 ± 0.1121
Our method distance1	0.2027 ± 0.0056	0.0786 ± 0.0041	0.2261 ± 0.0089
Our method distance3	0.2150 ± 0.0079	0.0908 ± 0.0029	0.2374 ± 0.0081
Our method distance4	**0.5997 ± 0.0252**	**0.5629 ± 0.0068**	**0.6420 ± 0.0120**

Table 7. Digits data set

Method	ACC	NMI	Purity
View1	0.5996 ± 0.0292	0.5868 ± 0.0089	0.6267 ± 0.0159
View2	0.4763 ± 0.0255	0.4679 ± 0.0112	0.5090 ± 0.0162
View3	0.6810 ± 0.0627	0.6891 ± 0.0241	0.7074 ± 0.0482
View4	0.6668 ± 0.0746	0.6417 ± 0.0314	0.7000 ± 0.0566
View5	0.4794 ± 0.0182	0.4586 ± 0.0114	0.5166 ± 0.0148
View6	0.4375 ± 0.0118	0.4914 ± 0.0038	0.4825 ± 0.0086
SMVBDC diatance1	0.7740 ± 0.0190	0.7375 ± 0.0124	0.7935 ± 0.0094
SMVBDC distance3	0.7057 ± 0.0476	0.6984 ± 0.0052	0.7495 ± 0.0295
SMVBDC distance4	0.4431 ± 0.0117	0.4941 ± 0.0034	0.4948 ± 0.0095
RMKMC	0.7437 ± 0.1832	0.7342 ± 0.1767	0.7626 ± 0.1844
Our method distance1	**0.8573 ± 0.0628**	**0.8143 ± 0.0259**	**0.8755 ± 0.0406**
Our method distance3	0.8063 ± 0.0006	0.7126 ± 0.0006	0.8063 ± 0.0006
Our method distance4	0.4485 ± 0.0073	0.4941 ± 0.0028	0.4919 ± 0.0057

Table 8. WebKB-texas data set

Method	ACC	NMI	Purity
View1	0.4016 ± 0.0160	0.1317 ± 0.0183	0.5928 ± 0.0168
View2	**0.4950 ± 0.0292**	0.1493 ± 0.0457	0.5882 ± 0.0341
View3	0.4214 ± 0.0157	0.1804 ± 0.0219	0.5925 ± 0.0134
View4	0.4536 ± 0.0539	**0.2092 ± 0.0355**	0.3581 ± 0.0350
SMVBDC diatance1	0.3925 ± 0.0247	0.1365 ± 0.0173	0.5186 ± 0.0283
SMVBDC distance3	0.4363 ± 0.0783	0.1438 ± 0.0063	0.5946 ± 0.0258
SMVBDC distance4	0.3775 ± 0.0167	0.1501 ± 0.0068	0.5903 ± 0.0207
RMKMC	0.2286 ± 0.0561	0.0101 ± 0.0032	0.3129 ± 0.0747
Our method distance1	0.4727 ± 0.0395	0.2036 ± 0.0221	**0.6561 ± 0.0200**
Our method distance3	0.4775 ± 0.0796	0.1589 ± 0.0196	0.6112 ± 0.0153
Our method distance4	0.4845 ± 0.0709	0.1457 ± 0.0039	0.6037 ± 0.0094

Table 9. WebKB-washington data set

Method	ACC	NMI	Purity
View1	0.4777 ± 0.0396	0.1714 ± 0.0356	0.6334 ± 0.0345
View2	0.4213 ± 0.0157	0.0860 ± 0.0112	0.4765 ± 0.0113
View3	0.4826 ± 0.0235	0.2053 ± 0.0084	0.6493 ± 0.0046
View4	0.5237 ± 0.0404	**0.3269 ± 0.0282**	0.7050 ± 0.0228
SMVBDC diatance1	0.3869 ± 0.0188	0.1578 ± 0.0316	0.5843 ± 0.0303
SMVBDC distance3	0.3539 ± 0.0258	0.0977 ± 0.0112	0.4913 ± 0.0082
SMVBDC distance4	0.3713 ± 0.0264	0.1177 ± 0.0292	0.5417 ± 0.0550
RMKMC	0.5183 ± 0.1369	0.2827 ± 0.0733	0.6418 ± 0.1556
Our method distance1	**0.5548 ± 0.0315**	0.3233 ± 0.0272	**0.7070 ± 0.0153**
Our method distance3	0.3891 ± 0.0386	0.1257 ± 0.0370	0.5357 ± 0.0449
Our method distance4	0.3939 ± 0.0161	0.1382 ± 0.0239	0.5770 ± 0.0450

Table 10. Cornell data set

Method	ACC	NMI	Purity
View1	0.3794 ± 0.0359	0.1433 ± 0.0330	0.5071 ± 0.0220
View2	**0.4654 ± 0.0147**	0.1112 ± 0.0201	0.4816 ± 0.0131
View3	0.3616 ± 0.0183	0.0994 ± 0.0141	0.4507 ± 0.0130
View4	0.4158 ± 0.0425	0.1820 ± 0.0501	0.5509 ± 0.0473
SMVBDC diatance1	0.3502 ± 0.0387	0.1316 ± 0.0288	0.4876 ± 0.0385
SMVBDC distance3	0.3733 ± 0.0480	0.1479 ± 0.0209	0.5153 ± 0.0233
SMVBDC distance4	0.3789 ± 0.0415	0.1455 ± 0.0128	0.5174 ± 0.0175
RMKMC	0.4242 ± 0.1056	0.2020 ± 0.0558	0.5168 ± 0.1310
Our method distance1	0.4446 ± 0.0286	**0.2377 ± 0.0288**	**0.5887 ± 0.0303**
Our method distance3	0.3954 ± 0.0443	0.1443 ± 0.0131	0.5154 ± 0.0161
Our method distance4	0.3949 ± 0.0503	0.1486 ± 0.0203	0.5159 ± 0.0220

Table 11. 20newsgroups data set

Method	ACC	NMI	Purity
View1	0.5136 ± 0.0723	0.2942 ± 0.0738	0.5218 ± 0.0688
View2	0.4574 ± 0.0657	0.2486 ± 0.0608	0.4746 ± 0.0537
View3	0.3163 ± 0.0432	0.1436 ± 0.0532	0.3349 ± 0.0463
SMVBDC diatance1	0.8146 ± 0.1124	0.7277 ± 0.1027	0.8282 ± 0.0988
SMVBDC distance3	0.6622 ± 0.1230	0.5850 ± 0.1685	0.6864 ± 0.1242
SMVBDC distance4	0.6622 ± 0.1173	0.5817 ± 0.1117	0.6850 ± 0.1034
RMKMC	0.4185 ± 0.1113	0.2068 ± 0.0697	0.4382 ± 0.1155
Our method distance1	**0.8240 ± 0.1008**	**0.7345 ± 0.1007**	**0.8326 ± 0.0912**
Our method distance3	0.7042 ± 0.1249	0.5985 ± 0.1269	0.7158 ± 0.1223
Our method distance4	0.4760 ± 0.0934	0.3002 ± 0.0690	0.4948 ± 0.0815

4   Conclusion

In this paper, we put forward a novel multi-view clustering algorithm based on Bregman divergences. It learns an appropriate weight for different views automatically by the constraints of diversity and low clustering loss. And the Bregman divergences control the loss caused by clustering efficiently during the processing of optimization. Moreover, the proposed algorithm preserves the simplicity and extensibility of the classical k-means algorithm, and it is a collection of multiple loss functions, which makes the algorithm more adaptable to multi-view data. Experimental results on nine multi-view data sets well demonstrate the effectiveness of the proposed method.

References

1. Banerjee, A., Merugu, S., Dhillon, I.S., Ghosh, J.: Clustering with Bregman divergences. J. Mach. Learn. Res. **6**(4), 1705–1749 (2005)
2. Bettoumi, S., Jlassi, C., Arous, N.: Collaborative multi-view K-means clustering. Soft. Comput. **1**, 1–9 (2017)
3. Bickel, S., Scheffer, T.: Multi-view clustering. In: IEEE International Conference on Data Mining, pp. 19–26 (2004)
4. Cai, D., He, X., Han, J., Huang, T.S.: Graph regularized nonnegative matrix factorization for data representation. IEEE Trans. Pattern Anal. Mach. Intell. **33**(8), 1548–1560 (2011)
5. Cai, X., Nie, F., Huang, H.: Multi-view K-means clustering on big data. In: International Joint Conference on Artificial Intelligence, pp. 2598–2604 (2013)
6. Chaudhuri, K., Kakade, S.M., Livescu, K., Sridharan, K.: Multi-view clustering via canonical correlation analysis. In: Proceedings of International Conference on Machine Learning, pp. 129–136 (2009)
7. Chen, X., Xu, X., Huang, J.Z., Ye, Y.: TW-k-means: automated two-level variable weighting clustering algorithm for multiview data. IEEE Trans. Knowl. Data Eng. **25**(4), 932–944 (2013)
8. Coates, A., Ng, A.Y.: Learning feature representations with K-means. In: Montavon, G., Orr, G.B., Müller, K.-R. (eds.) Neural Networks: Tricks of the Trade. LNCS, vol. 7700, 2nd edn, pp. 561–580. Springer, Heidelberg (2012). https://doi.org/10.1007/978-3-642-35289-8_30
9. Du, L., et al.: Robust multiple kernel K-means using L21-norm (2015)
10. Gong, X., Wang, F., Huang, L.: Weighted NMF-based multiple sparse views clustering for web items. In: Kim, J., Shim, K., Cao, L., Lee, J.-G., Lin, X., Moon, Y.-S. (eds.) PAKDD 2017 Part II. LNCS (LNAI), vol. 10235, pp. 416–428. Springer, Cham (2017). https://doi.org/10.1007/978-3-319-57529-2_33
11. Jiang, B., Qiu, F., Wang, L.: Multi-view clustering via simultaneous weighting on views and features. Appl. Soft Comput. **47**, 304–315 (2016)
12. Lee, D.D., Seung, H.S.: Algorithms for non-negative matrix factorization. In: International Conference on Neural Information Processing Systems, pp. 535–541 (2000)
13. Liu, H., Wu, J., Liu, T., Tao, D., Fu, Y.: Spectral ensemble clustering via weighted k-means: theoretical and practical evidence. IEEE Trans. Knowl. Data Eng. **29**(5), 1129–1143 (2017)
14. Liu, J., Wang, C., Gao, J., Han, J.: Multi-view clustering via joint nonnegative matrix factorization (2013)
15. Pu, J., Zhang, Q., Zhang, L., Du, B., You, J.: Multiview clustering based on robust and regularized matrix approximation. In: International Conference on Pattern Recognition, pp. 2550–2555 (2017)
16. Qian, Y., Li, F., Liang, J., Liu, B., Dang, C.: Space structure and clustering of categorical data. IEEE Trans. Neural Netw. Learn. Syst. **27**(10), 2047 (2016)
17. Tzortzis, G., Likas, A.: Kernel-based weighted multi-view clustering. In: IEEE International Conference on Data Mining, pp. 675–684 (2012)
18. Wang, F., Wang, X., Li, T.: Generalized cluster aggregation. In: IJCAI 2009, Proceedings of the International Joint Conference on Artificial Intelligence, Pasadena, California, USA, pp. 1279–1284, July 2009
19. Wang, H., Nie, F., Huang, H.: Multi-view clustering and feature learning via structured sparsity. In: International Conference on Machine Learning, pp. 352–360 (2013)

20. Wang, H., Yang, Y., Li, T.: Multi-view clustering via concept factorization with local manifold regularization. In: IEEE International Conference on Data Mining (2017)
21. Wang, J., Tian, F., Yu, H., Liu, C.H., Zhan, K., Wang, X.: Diverse non-negative matrix factorization for multiview data representation. IEEE Trans. Cybern. **1**(99), 1–13 (2017)
22. Xia, R., Pan, Y., Du, L., Yin, J.: Robust multi-view spectral clustering via low-rank and sparse decomposition. In: Twenty-Eighth AAAI Conference on Artificial Intelligence, pp. 2149–2155 (2014)
23. Xu, J., Han, J., Nie, F., Li, X.: Re-weighted discriminatively embedded K-means for multi-view clustering. IEEE Trans. Image Process. A Publ. IEEE Sig. Process. Soc. **26**(6), 3016–3027 (2017)
24. Xu, J., Han, J., Nie, F.: Discriminatively embedded K-means for multi-view clustering. In: Computer Vision and Pattern Recognition, pp. 5356–5364 (2016)
25. Yong, J.L., Grauman, K.: Foreground focus: unsupervised learning from partially matching images. Int. J. Comput. Vis. **85**(2), 143–166 (2009)
26. Yu, H., Lian, Y., Li, S., Chen, J.X.: View-weighted multi-view K-means clustering. In: Lintas, A., Rovetta, S., Verschure, P.F.M.J., Villa, A.E.P. (eds.) ICANN 2017 Part II. LNCS, vol. 10614, pp. 305–312. Springer, Cham (2017). https://doi.org/10.1007/978-3-319-68612-7_35
27. Zhan, K., Shi, J., Wang, J., Tian, F.: Graph-regularized concept factorization for multi-view document clustering. J. Vis. Commun. Image Represent. **48**, 4111–418 (2017)
28. Zhang, P., Yang, Y., Peng, B., He, M.: Multi-view clustering algorithm based on variable weight and MKL (2017)
29. Zhang, X., Wang, Z., Zong, L., Yu, H.: Multi-view clustering via graph regularized symmetric nonnegative matrix factorization. In: IEEE International Conference on Cloud Computing and Big Data Analysis, pp. 109–114 (2016)
30. Zong, L., Zhang, X., Zhao, L., Yu, H., Zhao, Q.: Multi-view clustering via multi-manifold regularized non-negative matrix factorization. Neural Netw. **88**, 74–89 (2017)

Graph-Based and Semi-supervised
Learning

Learning Safe Graph Construction
from Multiple Graphs

De-Ming Liang[1,2] and Yu-Feng Li[1,2(✉)]

[1] National Key Laboratory for Novel Software Technology, Nanjing University,
Nanjing 210023, China
{liangdm,liyf}@lamda.nju.edu.cn
[2] Collaborative Innovation Center of Novel Software Technology
and Industrialization, Nanjing 210023, China

Abstract. Graph-based method is one important paradigm of semi-supervised learning (SSL). Its learning performance typically relies on excellent graph construction which, however, remains challenging for general cases. What is more serious, constructing graph improperly may even deteriorate performance, which means its performance is worse than that of its supervised counterpart with only labeled data. For this reason, we consider learning a safe graph construction for graph-based SSL in this work such that its performance will not significantly perform worse than its supervised counterpart. Our basic idea is that, given a data distribution, there often exist some dense areas which are robust to graph construction. We then propose to combine trustable subgraphs in these areas from a set of candidate graphs to derive a safe graph, which remains to be a convex problem. Experimental results on a number of datasets show that our proposal is able to effectively avoid performance degeneration compared with many graph-based SSL methods.

Keywords: Safe · Graph construction · Semi-supervised learning

1 Introduction

Weakly supervised learning [31] is a core area in machine learning, among which semi-supervised learning [8,30] is the representative problem. It aims to improve learning performance via the usage of unlabeled data. During the past decades, extensive SSL studies have been presented, among which one popular paradigm is known as Graph-based SSL (GSSL) [5,27,28]. It is built on smooth assumption [8], i.e., similar instances share similar labels. Technically, it constructs a graph to encode the similarities between labeled and unlabeled data, and then learns a label assignment for unlabeled data in the goal that the inconsistency with respect to the constructed graph is minimized. GSSL is an SSL extension of classic supervised nearest neighbor method [11] and now is found useful for many diverse applications [19].

Nowadays it is widely accepted that the key for the success of GSSL is to construct an excellent graph for given training data, rather than designing various

© Springer Nature Singapore Pte Ltd. 2018
Z.-H. Zhou et al. (Eds.): ICAI 2018, CCIS 888, pp. 41–54, 2018.
https://doi.org/10.1007/978-981-13-2122-1_4

learning or optimization algorithms [3,10,24,30]. For this reason, many efforts have been devoted to graph construction during the decades, e.g., [7,10,23]. Generally, excellent graph construction remains challenging or an open problem, especially when domain knowledge is scarce or insufficient to afford a reliable graph construction.

Beyond constructing excellent graphs, one another or more serious issue is that constructing graph improperly may even deteriorate performance, which means its performance is worse than that of its supervised counterpart (supervised nearest neighbor method) with only labeled data [3,10,27,30]. These phenomena clearly conflicts with the original intention of GSSL. They will also encumber the deployment of GSSL in reality, because the GSSL users typically expect that employing GSSL methods should not be worse than direct supervised nearest neighbor methods. Therefore, it is highly desirable to derive safe graph constructions, which would not be outperformed by its supervised counterpart.

In order to tend this goal, in this work, we present a new GSSL method named SaGraph (Safe GRAPH). The basic intuition for our proposal aais that, given a data distribution, there often exist some dense areas which are robust or insensitive to graph construction. We refer to these areas as *safe* areas. For the cases where domain knowledge is scarce or insufficient to construct an excellent graph, one may be more reliable to construct a graph from the data in safe areas, so as to avoid the risk caused by an improper graph construction.

Based on this intuition, SaGraph proposes to exploit a set of candidate graphs and combines their trustable subgraphs in safe areas to derive a safe graph. To locate the safe areas, SaGraph optimizes the worst nearest neighbor error on each training instance according to a set of candidate graphs, and then treats the unlabeled data with the smallest nearest neighbor error (which implies that the prediction on these unlabeled data is not sensitive to graph construction) as the data in safe areas. The final optimization remains to be a convex problem. Experimental results on a number of datasets demonstrate that our proposal clearly improves the safeness of GSSL compared to many state-of-the-art methods.

In the following, we first introduce related work and then present our proposal. After that, we give the experimental justification and finally we conclude this work.

2   Related Work

This work is related to two branches of studies. One is GSSL and the other is safe SSL. In the aspect of GSSL, considerable attention has been paid since it was proposed, which can be separated into two categories. The first one works on various optimization algorithms, e.g. [4,5,10,11,27,28] and the second works on graph construction, e.g. [7,10,24]. There are also approaches, e.g., [1,29] proposed to optimize graph construction as well as label assignment simultaneously. It is notable that, as the deepening of research, graph construction is realized to be more important than the concrete optimization algorithms [3,10,24,30]. Nevertheless, generally, excellent graph construction remains challenging for GSSL.

Particularly, the research on explicitly constructing safe graph, to our best knowledge, has not been thoroughly studied yet.

In the aspect of safe SSL, this line of research is raised in very recent. [13, 14] is one early work to build safe semi-supervised SVMs. They optimize the worst-case performance gain given a set of candidate low-density separators, showing that the proposed S4VM (Safe Semi-Supervised SVM) is probably safe given that low-density assumption [8] holds. Later, a modified cluster assumption is proposed by Wang et al. [25] to safely utilize the unlabeled data. Krijthe and Loog [12] present to build a safe inductive semi-supervised classifier, which learns a projection of a supervised least square classifier from all possible semi-supervised least square ones. Apart from least square loss, a safe method for complex performance measures such as Top-k precision, F_β score or AUC is studied in [16]. Recently, Balsubramani and Freund [2] propose to learn a robust and high accurate prediction given that the ground-truth label assignment is restricted to one specific candidate set. Li et al. [15] study the quality of graph via a large margin principle and empirically achieve promising performance, while safe graph construction remains an open problem for GSSL. Besides, Wei et al. [26] study safe multi-label learning with weakly labeled data. Niu et al., [21] give a theoretical study about when positive unlabeled learning outperforms positive negative learning. Li et al. [17] cast the safe semi-supervised regression problem as a geometric projection issue with an efficient algorithm. Most recently, a general formulation for safe weakly supervised learning is proposed [9]. In this work, we consider a new scenario of safe SSL, i.e., safe graph construction that has not been studied.

3 The Proposed Method

In this section, we first briefly introduce a background for GSSL, and then present our idea and problem formulation, finally we derive its connection to safe graph construction and the learning algorithm.

3.1 Brief Background of GSSL

In SSL, we have l labeled instances $\{\mathbf{x}_i, \mathbf{y}_i\}_{i=1}^{l}$ and u unlabeled instances $\{\mathbf{x}_j\}_{j=l+1}^{l+u}$ ($l \ll u$). $\mathbf{y} = [y^1, \ldots, y^c] \in \{0, 1\}^c$ is the output label vector for input instance $\mathbf{x}$. c denotes the total number of classes, where each instance belongs to one class, i.e., $\sum_{h=1}^{c} y^h = 1$.

For GSSL, a graph $G = (\mathcal{V}, \mathcal{E}, \mathbf{W})$ is constructed for both the labeled and unlabeled data. Specifically, $\mathcal{V}$ is a set of $l + u$ nodes each corresponds to one instance. $\mathcal{E}$ is a set of undirected edges between node pairs. $\mathbf{W} \in \mathbb{R}^{(l+u) \times (l+u)}$ is a nonnegative and symmetric adjacency weighted matrix associating with $\mathcal{E}$ in G, i.e., the weight w_{ij} on the edge $e_{ij} \in \mathcal{E}$ reflects the similarity between $\mathbf{x}_i$ and $\mathbf{x}_j$. The larger the value w_{ij} is, the more similar $\mathbf{x}_i$ and $\mathbf{x}_j$ are. The goal of GSSL is to learn a label assignment $\{\tilde{\mathbf{y}}_j\}_{j=1}^{l+u}$ for training data such that the label inconsistency w.r.t. graph G is minimized. It is cast as the following optimization.

$$\min_{\{\tilde{\mathbf{y}}_j\}_{j=1}^{l+u}} \sum_{e_{ij} \in \mathcal{E}} w_{ij} \|\tilde{\mathbf{y}}_i - \tilde{\mathbf{y}}_j\|^2$$

$$\text{s.t.} \quad \tilde{\mathbf{y}}_j \in [0,1]^c, \sum_{h=1}^{c} \tilde{y}_j^h = 1, \forall j = l+1,\dots,l+u.$$

$$\tilde{\mathbf{y}}_i = \mathbf{y}_i, \forall i = 1,\dots,l. \tag{1}$$

It is worth noting that as stated in [11], the objective of GSSL (i.e., Eq. (1)) is a tight convex relaxation of supervised nearest neighbor error on training data. In other words, GSSL is no more than an SSL extension of classic supervised nearest neighbor algorithms for unlabeled data.

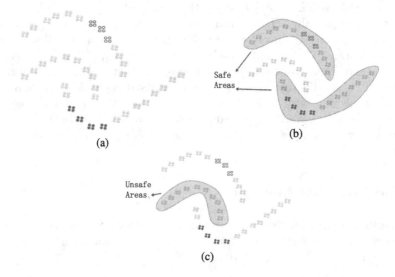

Fig. 1. Illustration for safe areas. (a) Labeled data (empty and filled circles) and unlabeled data (gray points). Given a data distribution, there exist some *safe* areas (b) that are robust to graph construction, and some unsafe areas (c) that are highly sensitive to graph construction.

3.2   Problem Formulation

Conventional GSSL methods typically aim to derive a good performance through a good graph construction. However, as mentioned previously, an inappropriate graph construction will cause GSSL to degenerate performance. To alleviate such a challenging problem, in this work we propose to learn a safe graph construction and present SAGRAPH. Unlike many GSSL methods which are developed on a certain graph, SAGRAPH considers to use a set of candidate graphs and exploits their trustable subgraphs to avoid performance decrease caused by improper graphs.

Figure 1 illustrates the intuition of SAGRAPH. Given several labeled data and a large amount of unlabeled data, there often exist some dense areas of data distribution which are robust or insensitive to graph construction. We refer to these areas as *safe* areas and the other areas as unsafe ones. Without sufficient domain knowledge to construct an excellent graph, one should only exploit the data as well as their subgraphs in safe areas to help improve the performance, and do not use the high risky data in unsafe areas.

The key is to locate safe areas. Remind that, according to the properties of safe areas, the data within safe areas should have small nearest neighbor errors with respect to multiple graphs. This motivates us the formulation of safe graph construction. Specifically, let $\{G_t = (\mathcal{V}, \mathcal{E}_t, \mathbf{W}_t)\}_{t=1}^{T}$ denote a set of candidate graphs, where T is the number of graphs. Let $er(\tilde{\mathbf{y}}_j)^{G_t}$ denote the nearest neighbor error of training data $\mathbf{x}_j$ on graph G_t, where it is defined as follows [11],

$$er(\tilde{\mathbf{y}}_j)^{G_t} = \delta(p_j^{G_t} \neq \arg\max_{h \in \{1,...,c\}} \tilde{y}_j^h)$$

δ is an indicator function. $p_j^{G_t} = \arg\max_{h \in \{1,...,c\}} \bar{p}_j^h$, where $\bar{\mathbf{p}}_j = [p_j^1, \ldots, p_j^c]$ is the prediction of $\mathbf{x}_j$ via classic nearest neighbor algorithm, i.e.,

$$\bar{\mathbf{p}}_j = \frac{\sum_{k:e_{jk} \in \mathcal{E}_t} \tilde{\mathbf{y}}_k w_{j,k}}{\sum_{k:e_{jk} \in \mathcal{E}_t} w_{jk}}$$

We then aim to locate data in safe areas. Specifically, given a label assignment $\tilde{\mathbf{Y}} = [\tilde{\mathbf{y}}_1, \tilde{\mathbf{y}}_2, \ldots, \tilde{\mathbf{y}}_{l+u}]$ for training data, the worst or equivalently maximal nearest neighbor error of training data $\mathbf{x}_j$ is,

$$\max_{t=1,...,T}\{er(\tilde{\mathbf{y}}_j)^{G_t}\}.$$

We then have our formulation for SAGRAPH as follows,

$$\min_{\tilde{\mathbf{Y}}} \sum_{j=l+1}^{l+u} \max_{t=1,...,T}\{er(\tilde{\mathbf{y}}_j)^{G_t}\}$$
$$\text{s.t. } \tilde{\mathbf{y}}_i = \mathbf{y}_i, \forall i = 1, \ldots, l,$$
$$\tilde{\mathbf{y}}_j \in [0,1]^c, \forall j = l+1, \ldots, l+u.$$
$$\sum_{h=1}^{c} \tilde{y}_i^h = 1, \forall i = 1, \ldots, l+u.$$
$$\sum_{j=l+1}^{l+u} \tilde{\mathbf{y}}_j/u = \sum_{i=1}^{l} \tilde{\mathbf{y}}_i/l. \tag{2}$$

The last equality is balance constraint [11] in order to avoid trivial solutions which predict all training data to one class. However, the optimization in Eq. (2) is non-trivial, as the form of $er(\tilde{\mathbf{y}}_j)^{G_t}$ is non-continuous and non-convex. Inspired by the way in many GSSL studies [11], a tight convex upper bound of $er(\tilde{\mathbf{y}}_j)^{G_t}$ is optimized alternatively, and we have the following convex form for SAGRAPH,

$$\min_{\tilde{\mathbf{Y}} \in \Omega} \sum_{j=l+1}^{l+u} \max_{t=1,...,T}\{\sum_{i:e_{ij} \in \mathcal{E}_t} \tilde{w}_{ij}^t(\|\tilde{\mathbf{y}}_i - \tilde{\mathbf{y}}_j\|^2)\} \tag{3}$$

Algorithm 1. The SAGRAPH Method

Input: Few label instances $\{\mathbf{x}_i, \mathbf{y}_i\}_{i=1}^{l}$, a large amount of unlabeled instances $\{\mathbf{x}_j\}_{j=l+1}^{l+u}$, a set of candidate graphs $\{G_t = (\mathcal{V}, \mathcal{E}_t, \mathbf{W}_t)\}_{t=1}^{T}$, a parameter ϵ;

Output: A label assignment for labeled and unlabeled data $\hat{\mathbf{Y}} = [\hat{\mathbf{y}}_1, \ldots, \hat{\mathbf{y}}_{l+u}]$;

  1: Address the convex optimization problem in Eq. (5)

  2: Denote $\tilde{\mathbf{Y}}^* = [\tilde{\mathbf{y}}_1^*, \ldots, \tilde{\mathbf{y}}_{l+u}^*]$ as the optimal solution;

  3: $\tilde{\mathbf{y}}_j^* = \tilde{\mathbf{y}}_j^* - \sum_{i=1}^{l} \tilde{\mathbf{y}}_i/l$, $u_j = \arg\max_{h=1,\ldots,c} \tilde{y}_j^{*,h}$, $\bar{u}_j = \arg\max_{h\neq u_j} \tilde{y}_j^{*,h}$, $\forall j = l+1, \ldots, l+u$;

  4: For $\mathbf{x}_j \in \mathcal{B} = \{\mathbf{x}_k | \tilde{y}_k^{*,u_k} - \tilde{y}_k^{*,\bar{u}_k} \geq \epsilon, \forall j = l+1, \ldots, l+u\}$, predict $\mathbf{x}_j$ to class u_j; Otherwise, predict $\mathbf{x}_j$ with a direct supervised counterpart;

  5: Return $\hat{\mathbf{Y}} = [\hat{\mathbf{y}}_1, \ldots, \hat{\mathbf{y}}_{l+u}]$;

---

where Ω refers to the feasible set of $\tilde{\mathbf{Y}}$, i.e.,

$$\Omega = \{\tilde{\mathbf{Y}} | \tilde{\mathbf{y}}_j = \mathbf{y}_j, j = 1, \ldots, l;$$
$$\tilde{\mathbf{y}}_j \in [0,1]^c, j = l+1, \ldots, l+u;$$
$$\sum_{h=1}^{c} \tilde{y}_i^h = 1, i = 1, \ldots, l+u;$$
$$\sum_{j=l+1}^{l+u} \tilde{\mathbf{y}}_j/u = \sum_{i=1}^{l} \tilde{\mathbf{y}}_i/l\}.$$

As Eq. (3) shows, SAGRAPH considers the worst nearest neighbor error, rather than direct nearest neighbor error in typical GSSL methods, tending to derive robust predictions.

3.3 Connection to Safe Graph Construction and Learning Algorithm

In this section we first present that Eq. (3) leads to a safe graph construction and then present the algorithm for Eq. (3). Specifically, by introducing additional variables $\boldsymbol{\theta} = [\theta_{l+1}, \ldots, \theta_{l+u}]$ for each unlabeled data, Eq. (3) can be written in an equivalent way as Eq. (4),

$$\min_{\tilde{\mathbf{Y}} \in \Omega, \boldsymbol{\theta}} \sum_{j=l+1}^{l+u} \theta_j \qquad (4)$$
$$\text{s.t} \quad \theta_j \geq \sum_{i:e_{ij} \in \mathcal{E}_t} \tilde{w}_{ij}^t (\|\tilde{\mathbf{y}}_i - \tilde{\mathbf{y}}_j\|^2), \forall t = 1, \ldots, T.$$

By further introducing dual variables $\boldsymbol{\alpha} = [\alpha_{1,1}, \ldots, \alpha_{1,T}, \alpha_{2,1}, \ldots, \alpha_{u,T}]$ each corresponds to one constraint in Eq. (4), and setting the derivatives with respect to $\boldsymbol{\theta}$ to zero, Eq. (4) can be further rewritten as Eq. (5),

$$\min_{\tilde{\mathbf{Y}} \in \Omega} f(\tilde{\mathbf{Y}}) \qquad (5)$$

Table 1. Statistics of data sets.

Data	# Dim.	# Inst.	# Clas.	Data	# Dim.	# Inst.	# Clas.	Data	# Dim.	# Inst.	# Clas.
austra	15	690	2	coil	241	1,500	2	house	16	232	2
bci	117	400	2	digit1	241	1,500	2	house-votes	16	435	2
usps	256	7,291	10	dna	180	2,000	3	ionosphere	33	351	2
breastw	9	683	2	glass	9	214	6	liverDisorders	6	345	2
clean1	166	476	2	heart	9	270	2	isolet	51	600	2
iris	4	150	3	wine	13	178	3	vehicle	18	846	4
wdbc	14	569	2	text	11,960	1,500	2	-	–	-	-

where

$$f(\tilde{\mathbf{Y}}) \triangleq \max_{\alpha \in \mathcal{M}} \sum_{j=l+1}^{l+u} \sum_{t=1}^{T} \alpha_{j-l,t} \sum_{i:e_{ij} \in \mathcal{E}_t} \tilde{w}_{ij}^t \big(\|\tilde{\mathbf{y}}_i - \tilde{\mathbf{y}}_j\|^2 \big)$$

$$\mathcal{M} = \{\alpha | \alpha \geq 0; \sum_{t=1}^{T} \alpha_{j-l,t} = 1, \forall j = l+1, \ldots, l+u\}.$$

We now show that Eq. (5) can be regarded as conventional GSSL methods on a "safe" graph. Specifically, let α^* denote the optimal solution to Eq. (5). We define a new graph $G^* = \{\mathcal{V}, \mathcal{E}^*, \mathbf{W}^*\}$ as follows,

- $e_{ij} \in \mathcal{E}^*$ if and only if $\exists t, \alpha_{j-l,t}^* > 0$ & $e_{ij} \in \mathcal{E}_t$
- $w_{ij}^* = \sum_{t=1}^{T} \alpha_{j-l,t}^* \tilde{w}_{ij}^t$ if and only if $e_{ij} \in \mathcal{E}^*$

Refer to the definition of $f(\tilde{\mathbf{Y}})$ and according to the property for optimal solution [20], we can rewrite Eq. (5) in an equivalent way as Eq. (6),

$$\min_{\tilde{\mathbf{Y}} \in \Omega} \sum_{e_{ij} \in \mathcal{E}^*} w_{ij}^* \|\tilde{\mathbf{y}}_i - \tilde{\mathbf{y}}_j\|^2. \tag{6}$$

By comparing Eq. (6) with Eq. (1), it is not hard to find that Eq. (6) (or equivalently Eq. (5)) is no more than a GSSL method based on a "safe" graph $G^* = \{\mathcal{V}, \mathcal{E}^*, \mathbf{W}^*\}$.

As for solving Eq. (5), it is easy to find that Eq. (5) is a convex optimization, and can be addressed via state-of-the-art efficient optimization technique. Further note that the objective of Eq. 5 is a convex yet non-smooth, and thus gradient-based method is not applicable. In this paper we adopt the projected subgradient optimal method [20] for the solving of Eq. (5), which was shown as the fastest first-order method with optimal convergence rate. After solving Eq. (5), the nearest neighbor errors within the safe areas are small to various graphs and in contrast, the nearest neighbor errors within unsafe areas are diverse. Therefore, the unlabeled data with high confident label assignment are realized as the data in safe areas, while the rest ones are the data in unsafe areas. Algorithm 1 summarizes the pseudo codes of SaGraph.

3.4   Efficient SGD Optimization

To optimize Eq. (5), it is a saddle-point convex-concave optimization problem. Direct subgradient approaches are generally slow due to a poor convergence rate [20]. We adopt efficient stochastic gradient descent (SGD) proposed in [18] to solve Eq. (5). Specifically, the key for SGD is to derive an unbiased estimator of full gradient. Let

$$g_j(\tilde{\mathbf{Y}}) = \max_{\alpha \in \mathcal{M}} \sum_{t=1}^{T} \alpha_{j-l,t} \sum_{i:e_{ij} \in \mathcal{E}_t} \tilde{w}_{ij}^t \left(\|\tilde{\mathbf{y}}_i - \tilde{\mathbf{y}}_j\|^2 \right).$$

Therefore, $g_j(\tilde{\mathbf{Y}})$ is the value of worst-case nearest neighbor error at the $(j-l)$-th unlabeled instance, and $f(\tilde{\mathbf{Y}}) = \sum_{j=l+1}^{l+u} g_j(\tilde{\mathbf{Y}})$ is the total error. In SGD, the full gradient of $f(\tilde{\mathbf{Y}})$ is approximated by a gradient at a singe instance as it is unbiased to full gradient:

$$\tilde{\mathbf{Y}}_{r+1} := Proj_\Omega(\tilde{\mathbf{Y}}_r - \eta_r \nabla g_j(\tilde{\mathbf{Y}}_r)), \ \forall j = l+1, \ldots, l+u \tag{7}$$

$\tilde{\mathbf{Y}}_r$ is the solution in r-th iteration, η_r is the step size and $Proj_\Omega(\mathbf{Y})$ refers to the projection of $\mathbf{Y}$ onto Ω. Through SGD, one can update $\tilde{\mathbf{Y}}$ without calculating the full gradient. Further note that the calculation of $\nabla g_j(\tilde{\mathbf{Y}})$ is cheap as it is only related to a small piece of nodes/instances (i.e., the set of nearest neighbors for each unlabeled instance) and can be computed in a very efficient manner. Specifically, let $h_{j,t}(\tilde{\mathbf{Y}}) = \sum_{i:e_{ij} \in \mathcal{E}_t} \tilde{w}_{ij}^t \left(\|\tilde{\mathbf{y}}_i - \tilde{\mathbf{y}}_j\|^2 \right)$ and $mt = \arg\max_{t=1,\ldots,T} h_{j,t}(\tilde{\mathbf{Y}}_r)$. According to the rule of subgradient [20], $\nabla g_j(\tilde{\mathbf{Y}}_r) = \nabla h_{j,mt}(\tilde{\mathbf{Y}}_r)$ is cheap to compute. It is not hard to have the following convergence results by adopting the proof from [6].

Theorem 1. *Let B be the norm of $\tilde{\mathbf{Y}}$ (i.e., $\|\tilde{\mathbf{Y}}\|_2 \leq B$), and G be the upper bound of $\nabla g_j(\tilde{\mathbf{Y}}_r)$ (i.e., $\|\nabla g_j(\tilde{\mathbf{Y}}_r)\|_2 \leq G$). By setting $\eta_r = \frac{B/G}{\sqrt{r}}$, one can have*

$$\mathbb{E}[f(\tilde{\mathbf{Y}}_r)] - f(\tilde{\mathbf{Y}}^*) \leq O(\frac{GB}{\sqrt{r}})$$

Further more, it is easy to show that

Proposition 1. *$B = O((l+u)c)$ and G is a small constant.*

In practice the convergence rate is much better than the theoretical bound. The algorithm often obtains a quite good approximate solution by passing the unlabeled instances in less than five times. In other words, linear running time $O(|E|)$ is usually sufficient to obtain a good empirical result, where $|E|$ is the maximal number of edges among candidate graphs.

After solving Eq. (5), the nearest neighbor errors within the safe areas are small to various graphs and in contrast, the nearest neighbor errors within unsafe areas are diverse. Therefore, the unlabeled data with high confident label assignment are realized as the data in safe areas, while the rest are the data in unsafe ones. Algorithm 1 summarizes the pseudo codes of SaGraph.

4   Experiments

We evaluate our SAGRAPH method on a broad range of data sets in this section. The size of dataset ranges from 150 to over 7,000, the dimension of instance ranges from 4 to more than 10,000, and the tasks cover both binary and multi-class classification. Table 1 summarizes the detail statistics of data sets.

4.1   Experimental Setup

The proposed method is compared with the following methods including one baseline supervised method and four classic GSSL methods.

- **k-NN**: supervised k nearest neighbor method, which serves as the baseline method. The key of this work is to show whether GSSL methods could always outperform such a baseline method or not.
- **GFHF**: Gaussian Field Harmonic Function [28]. This method formulates GSSL as gaussian field and uses belief propagation technique for inference, and finally a closed-form solution is derived.
- **CMN**: Class Mass Normalization [28]. This method is a variant of GFHF. CMN further enforces that its prediction on unlabeled data fits a balance constraint, avoiding the cases that all predictions are categorized into one class.
- **LGC**: Local and Global Consistency [27]. This method is motived by the spirit of random walk. It adopts label propagation idea and formulates GSSL as a convex problem with a simple iterative algorithm.
- **SGT**: Spectral Graph Transduction [11]. This method revisits GSSL as an extension form of the nearest neighbor algorithm. It then formulates GSSL as regularized normalized cut form and provides an efficient algorithm via the spectral of Laplacian matrix.

We adopt the code shared in the websites[1][2] for the implementations of GFHF, CMN and SGT. LGC is implemented by ourselves. For LGC and SGT, the parameters are set to the ones recommended in the paper; For SAGRAPH, the parameter ϵ is fixed to 0.3 on all cases.

It is notable that for all GSSL methods, graph construction may exist the cases that some connected components do not contain any labeled data. In this case, inferring the label for the unlabeled instances in these connected components is infeasible. In this paper, we assign such unlabeled instances with the predictive results of supervised k-nearest neighbor method. Note that such a strategy have already improved the safeness of classic GSSL methods. For all the experiments, 10 instances are randomly selected as labeled data for binary classification and 20 instances are randomly selected as labeled data for multi-class classification. The rest are employed as unlabeled data. Experiments repeat for 20 times and average accuracies with standard deviations are reported.

---

[1] http://pages.cs.wisc.edu/~jerryzhu/pub/harmonic_function.m.

[2] http://sgt.joachims.org/.

Table 2. The accuracies (with standard derivations) of supervised 1 nearest neighbor (1NN) method and multiple GSSL methods. The number in bold (resp. underline) means that the performance is significantly better (resp. worse) than supervised 1NN method (paired t-test at 95% confidence level). The results of Win/Tie/Loss are also listed in the last row, and the methods with the fewest losses are highlighted.

Data	1NN	GFHF	CMN	LLGC	SGT	SaGraph
aust	63.4 ± 6.5	63.5 ± 6.6	58.8 ± 6.0	63.6 ± 6.5	63.5 ± 6.6	63.4 ± 6.6
bci	51.3 ± 2.8	51.3 ± 2.9	45.2 ± 2.5	51.3 ± 2.8	51.3 ± 2.8	51.2 ± 2.8
brea	93.3 ± 3.9	93.3 ± 4.1	78.3 ± 4.5	93.3 ± 4.1	93.3 ± 4.1	93.0 ± 7.3
clea	58.6 ± 5.1	58.7 ± 5.2	52.1 ± 5.7	58.7 ± 5.2	58.7 ± 5.2	58.4 ± 5.6
coil	58.1 ± 6.3	**58.2 ± 6.3**	56.3 ± 6.4	**58.2 ± 6.3**	**58.2 ± 6.3**	58.2 ± 6.5
digi	77.5 ± 5.7	77.5 ± 5.7	75.2 ± 5.7	77.5 ± 5.7	77.5 ± 5.7	78.5 ± 9.2
hear	71.1 ± 5.6	71.2 ± 5.7	60.0 ± 5.5	71.1 ± 5.7	71.2 ± 5.7	71.6 ± 5.6
hous	89.0 ± 2.3	87.9 ± 3.0	70.2 ± 4.1	87.9 ± 3.0	88.0 ± 2.8	85.0 ± 12.7
houv	86.6 ± 3.1	86.3 ± 3.0	72.7 ± 4.5	86.3 ± 3.0	86.3 ± 2.9	**87.3 ± 3.0**
iono	73.3 ± 6.2	75.8 ± 6.7	51.1 ± 6.1	75.7 ± 6.7	75.7 ± 6.7	73.4 ± 6.4
isol	91.3 ± 4.1	91.3 ± 4.0	84.4 ± 3.9	91.3 ± 4.0	91.3 ± 4.0	**94.1 ± 4.4**
live	52.9 ± 3.2	52.6 ± 3.2	46.1 ± 3.9	52.6 ± 3.2	52.6 ± 3.3	52.8 ± 3.3
text	59.8 ± 3.5	59.7 ± 3.5	57.2 ± 3.5	59.7 ± 3.5	59.7 ± 3.5	59.8 ± 3.4
wdbc	81.2 ± 5.6	81.3 ± 5.3	65.1 ± 5.2	81.3 ± 5.3	81.3 ± 5.3	81.2 ± 5.6
dna	54.8 ± 3.3	**57.4 ± 4.2**	**57.5 ± 3.4**	**57.1 ± 4.5**	-	54.8 ± 3.6
glas	55.9 ± 4.3	55.9 ± 3.9	55.9 ± 4.0	56.1 ± 3.9	-	56.2 ± 4.2
iris	94.7 ± 2.9	94.3 ± 3.2	94.3 ± 3.2	94.2 ± 3.2	-	94.8 ± 2.3
usps	64.4 ± 4.5	**64.5 ± 4.5**	64.5 ± 4.5	**64.5 ± 4.5**	-	**68.2 ± 5.4**
vehi	48.1 ± 4.3	48.2 ± 4.3	48.2 ± 4.3	48.2 ± 4.3	-	**48.6 ± 4.3**
wine	92.0 ± 1.8	90.5 ± 2.0	90.5 ± 2.0	90.5 ± 2.0	-	92.5 ± 1.5
GSSL methods against 1NN: win/tie/loss		3/15/2	2/3/15	3/15/2	1/12/1	**4/16/0**

4.2 Comparison Results

We first compare with supervised 1 nearest neighbor (1NN) algorithm. Compared GSSL methods employ 1NN graph as the graph construction. For SaGraph method, 1NN, 3NN and 5NN graphs are adopted as candidate graphs. Since SGT focuses on binary classification, it does not have results on multi-class data sets.

Table 2 shows the compared results. As can be seen, in terms of *win* counts, i.e., the times where GSSL method significantly outperforms supervised 1NN method, SaGraph obtains the highest times compared with classic GSSL methods. More importantly, as Table 2 shows, compared GSSL methods did will significantly decrease performance in some cases, while SaGraph does not suffer from such a deficiency on these data sets. CMN does not work well, mainly because the class mass normalization is particularly suitable for one or very few connected components, which may be not that cases when using 1NN graph.

Table 2 shows that compared GSSL methods only obtain comparative performance with 1NN method. One reason is that 1NN graph is not so powerful for GSSL methods. We then evaluate our method in comparison to supervised 5 nearest neighbor (1NN) algorithm. Compared GSSL methods all employ 5NN graph as the graph construction. Following the same setup as before, 1NN, 3NN

Table 3. The accuracies (with standard derivations) of 5NN method and multiple GSSL methods.

Data	5NN	GFHF	CMN	LLGC	SGT	SaGraph
aust	62.5 ± 5.2	59.8 ± 4.3	59.6 ± 4.5	59.6 ± 4.3	56.1 ± 6.1	62.8 ± 5.5
bci	50.3 ± 2.8	49.8 ± 2.1	50.0 ± 2.2	49.7 ± 1.9	49.7 ± 2.1	50.3 ± 2.4
brea	90.2 ± 3.4	**96.0 ± 0.9**	**96.0 ± 0.6**	**95.9 ± 0.7**	**96.6 ± 0.2**	94.4 ± 3.1
clea	53.4 ± 3.4	**57.6 ± 4.8**	**57.8 ± 4.8**	**57.5 ± 5.1**	55.1 ± 4.7	54.4 ± 4.4
coil	47.6 ± 5.5	**66.5 ± 6.0**	50.3 ± 9.9	**64.9 ± 7.2**	**65.5 ± 7.5**	**53.3 ± 8.2**
digi	69.2 ± 8.3	**88.2 ± 5.3**	87.4 ± 3.2	**90.8 ± 3.2**	**95.0 ± 1.8**	71.8 ± 9.2
hear	70.2 ± 5.6	67.2 ± 7.3	69.4 ± 6.7	61.3 ± 5.5	57.8 ± 5.8	71.6 ± 4.4
hous	89.7 ± 1.8	84.0 ± 8.6	88.3 ± 2.3	78.4 ± 11.4	89.5 ± 1.4	84.2 ± 13.0
houv	85.9 ± 4.9	84.1 ± 6.6	87.7 ± 1.7	82.7 ± 7.5	**88.3 ± 0.4**	84.6 ± 7.4
iono	66.6 ± 2.6	**75.2 ± 9.1**	**77.2 ± 6.6**	**73.2 ± 7.3**	68.3 ± 8.9	**70.1 ± 4.2**
isol	91.7 ± 4.0	**97.6 ± 1.3**	**98.3 ± 0.6**	**97.5 ± 1.3**	**98.3 ± 0.4**	**93.9 ± 4.6**
live	52.0 ± 4.5	51.4 ± 4.7	52.1 ± 3.4	52.4 ± 3.7	51.5 ± 3.3	52.0 ± 4.4
text	57.2 ± 3.4	51.8 ± 2.0	**64.2 ± 5.3**	56.4 ± 6.1	61.0 ± 11.0	**57.4 ± 3.4**
wdbc	77.6 ± 5.1	76.9 ± 8.2	79.8 ± 4.8	70.7 ± 5.1	**94.3 ± 0.5**	**78.2 ± 5.2**
dna	56.4 ± 4.5	52.5 ± 5.5	53.3 ± 0.4	53.3 ± 0.4	-	**56.6 ± 4.6**
glas	49.5 ± 5.3	**54.6 ± 3.8**	**54.5 ± 4.1**	52.0 ± 5.8	-	**52.7 ± 4.9**
iris	92.5 ± 2.6	**94.0 ± 3.5**	**95.0 ± 1.8**	92.6 ± 3.0	-	**93.5 ± 2.6**
usps	46.6 ± 4.5	**67.5 ± 7.4**	**79.7 ± 7.5**	**86.8 ± 4.9**	-	**49.3 ± 5.1**
vehi	38.5 ± 4.1	**50.4 ± 5.6**	**51.9 ± 5.0**	**49.9 ± 5.5**	-	**41.9 ± 5.2**
wine	93.7 ± 1.9	94.2 ± 1.8	94.5 ± 1.0	93.7 ± 3.0	-	93.9 ± 1.8
GSSL methods against 5NN: win/tie/loss		10/6/4	10/7/3	8/7/5	6/6/2	**13/7/0**

and 5NN graphs are adopted as candidate graphs for SaGraph. SGT works on binary classification and does not have results on multi-class data.

The comparison results are shown in Table 3. As can be seen, in the case of 5NN graph, compared GSSL methods obtain more aggressive results, e.g., CMN works much better because 5NN graph leads to much less connected component. Even in this situation, SaGraph still achieves the highest times in terms of *win* counts. More worth mentioning is that, all compared GSSL methods suffer much more serious issue on performance degeneration than that in Table 3. While the proposal SaGraph still does not decrease performance significantly in this situation.

We now summarize the results in Tables 2 and 3.

- Direct GSSL methods did decrease performance significantly in considerable cases, no matter for sparse (1NN) or denser (5NN) graphs. Our proposal SaGraph does not suffer from this issue in all the cases reported in Tables 2 and 3.
- SaGraph achieves the highest times in terms of *win* counts in both Tables 2 and 3, showing that SaGraph owns a good ability in performance gain.
- In the aspect for the comparison between 1NN and 5NN graphs, GSSL methods with 5NN graph obtain more aggressive results (especially for CMN method), however, they also suffer from a more serious issue on performance degeneration.

Table 4. Influence on the number of candidate graphs for the SAGRAPH method

Win Counts												
GFHF	CMN	LLGC	SGT	Number of candidate graphs in SAGRAPH								
				2	3	4	5	6	7	8	9	10
6	6	6	6	7	7	7	7	7	7	7	7	7

Loss Counts												
GFHF	CMN	LLGC	SGT	Number of candidate graphs in SAGRAPH								
				2	3	4	5	6	7	8	9	10
3	2	1	4	0	0	0	0	0	0	0	0	0

4.3   Influence on Candidate Graphs

The situation of candidate graphs is one factor of our proposal. We further study the influence on the candidate graphs for SAGRAPH. Specifically, we generate the candidate graphs as follows. For each training instance, the number of nearest neighbors is randomly picked up from 1 to 9 with uniform distribution. In this case, 1NN, 3NN and 5NN graphs are special cases of these candidate graphs. The number of candidate graphs are set from 2 to 10. The experiments are conducted for 20 times and the win/loss counts against supervised 5NN method on binary classifications are shown in Table 4. The win/loss counts of compared GSSL methods are also listed for comparison. Table 4 shows that, SAGRAPH consistently works quite well on the number of candidate graphs, i.e., competitive win counts and much fewer loss counts. One reason is that the proposal only exploits some reliable subgraphs of candidate ones, rather than full graphs.

5   Conclusion and Future Work

In this paper we propose to learn a safe graph for graph-based SSL (GSSL), which could always outperform its supervised counterpart, i.e., classic nearest neighbor method. This is motivated by a crucial issue of GSSL that GSSL with the use of inappropriate graph construction may cause serious performance degeneration which could be even worse than its supervised counterpart. To overcome this issue, in this work we present an SAGRAPH method. The basic idea is that given a data distribution, there often exist some dense areas (or safe areas) which are quite robust or less sensitive to graph construction. One should exploits the data and the subgraphs in safe areas to learn a safe graph. We then consequently formulate the above consideration as a convex optimization and connect it to safe graph construction. Empirical studies on a number of data sets verify that our proposal achieves promising performance on the safeness of GSSL.

In our work achieving safe graph construction requires additional costs, e.g., more running time or smaller performance gain in some cases. One reason in that safe SSL needs to always take the safeness of SSL into account, whereas previous SSL studies do not have to take such kind of consideration and behave to be more aggressive. In future, we will study scalable safe GSSL as well as some other effective safe SSL approaches such as incorporating specific domain knowledge, e.g., known laws of physics [22].

Acknowledgement. The authors want to thank the reviewers for their helpful comments. This research was supported by the National Natural Science Foundation of China (61772262) and the Fundamental Research Funds for the Central Universities (020214380044).

References

1. Argyriou, A., Herbster, M., Pontil, M.: Combining graph Laplacians for semi-supervised learning. In: Advances in Neural Information Processing Systems, Cambridge, MA, pp. 67–74 (2005)
2. Balsubramani, A., Freund, Y.: Optimally combining classifiers using unlabeled data. In: Proceedings of International Conference on Learning Theory, Paris, France, pp. 211–225 (2015)
3. Belkin, M., Niyogi, P.: Towards a theoretical foundation for Laplacian-based manifold methods. J. Comput. Syst. Sci. **74**(8), 1289–1308 (2008)
4. Belkin, M., Niyogi, P., Sindhwani, V.: Manifold regularization: a geometric framework for learning from labeled and unlabeled examples. J. Mach. Learn. Res. **7**, 2399–2434 (2006)
5. Blum, A., Chawla, S.: Learning from labeled and unlabeled data using graph mincuts. In: Proceedings of the 8th International Conference on Machine Learning, Williamstown, MA, pp. 19–26 (2001)
6. Bottou, L., Curtis, F.E., Nocedal, J.: Optimization methods for large-scale machine learning. arXiv preprint arXiv:1606.04838 (2016)
7. CarreiraPerpiñán, M.Á., Zemel, R.S.: Proximity graphs for clustering and manifold learning. In: Advances in Neural Information Processing Systems, Cambridge, MA, pp. 225–232 (2005)
8. Chapelle, O., Schölkopf, B., Zien, A. (eds.): Semi-Supervised Learning. MIT Press, Cambridge (2006)
9. Guo, L.-Z., Li, Y.-F.: A general formulation for safely exploiting weakly supervised data. In: Proceedings of the 32nd AAAI Conference on Artificial Intelligence, New Orleans, LA (2018)
10. Jebara, T., Wang, J., Chang, S.F.: Graph construction and b-matching for semi-supervised learning. In: Proceedings of the 26th International Conference on Machine Learning, Montreal, Canada, pp. 441–448 (2009)
11. Joachims, T.: Transductive learning via spectral graph partitioning. In: Proceedings of the 20th International Conference on Machine Learning, Washington, DC, pp. 290–297 (2003)
12. Krijthe, J.H., Loog, M.: Implicitly constrained semi-supervised least squares classification. In: Fromont, E., De Bie, T., van Leeuwen, M. (eds.) IDA 2015. LNCS, vol. 9385, pp. 158–169. Springer, Cham (2015). https://doi.org/10.1007/978-3-319-24465-5_14

13. Li, Y.-F., Zhou, Z.-H.: Towards making unlabeled data never hurt. In: Proceedings of the 28th International Conference on Machine Learning, Bellevue, WA, pp. 1081–1088 (2011)
14. Li, Y.-F., Zhou, Z.-H.: Towards making unlabeled data never hurt. IEEE Trans. Pattern Anal. Mach. Intell. **37**(1), 175–188 (2015)
15. Li, Y.-F., Wang, S.-B., Zhou, Z.-H.: Graph quality judgement: a large margin expedition. In: Proceedings of the 25th International Joint Conference on Artificial Intelligence, New York, NY, pp. 1725–1731 (2016)
16. Li, Y.-F., Kwok, J., Zhou, Z.-H.: Towards safe semi-supervised learning for multivariate performance measures. In: Proceedings of the 30th AAAI Conference on Artificial Intelligence, Phoenix, AZ, pp. 1816–1822 (2016)
17. Li, Y.-F., Zha, H.-W., Zhou, Z.-H.: Learning safe prediction for semi-supervised regression. In: Proceedings of the 31st AAAI Conference on Artificial Intelligence, San Francisco, CA, pp. 2217–2223 (2017)
18. Liang, D.-M., Li, Y.-F.: Lightweight label propagation for large-scale network data. In: Proceedings of the 27th International Joint Conference on Artificial Intelligence, Stockholm, Sweden (2018)
19. Liu, W., Wang, J., Chang, S.F.: Robust and scalable graph-based semisupervised learning. Proc. IEEE **100**(9), 2624–2638 (2012)
20. Nesterov, Y.: Introductory Lectures on Convex Optimization. A Basic Course. Springer, Heidelberg (2003). https://doi.org/10.1007/978-1-4419-8853-9
21. Niu, G., du Plessis, M.C., Sakai, T., Ma, Y., Sugiyama, M.: Theoretical comparisons of positive-unlabeled learning against positive-negative learning. In: Advances in Neural Information Processing Systems, Barcelona, Spain, pp. 1199–1207 (2016)
22. Stewart, R., Ermon, S.: Label-free supervision of neural networks with physics and domain knowledge. In: Proceedings of 31th AAAI Conference on Artificial Intelligence, San Francisco, CA (2017)
23. Wang, F., Zhang, C.: Label propagation through linear neighborhoods. In: Proceedings of the 23rd International Conference on Machine Learning, Pittsburgh, PA, pp. 985–992 (2006)
24. Wang, F., Zhang, C.: Label propagation through linear neighborhoods. IEEE Trans. Knowl. Data Eng. **20**(1), 55–67 (2008)
25. Wang, Y.-Y., Chen, S.-C., Zhou, Z.-H.: New semi-supervised classification method based on modified cluster assumption. IEEE Trans. Neural Netw. Learn. Syst. **23**(5), 689–702 (2012)
26. Wei, T., Guo, L.-Z., Li, Y.-F., Gao, W.: Learning safe multi-label prediction for weakly labeled data. Mach. Learn. **107**(4), 703–725 (2018)
27. Zhou, D., Bousquet, O., Navin Lal, T., Weston, J., Schölkopf, B.: Learning with local and global consistency. In: Advances in Neural Information Processing Systems, pp. 595–602. MIT Press, Cambridge (2004)
28. Zhu, X., Ghahramani, Z., Lafferty, J.: Semi-supervised learning using Gaussian fields and harmonic functions. In: Proceedings of the 20th International Conference on Machine learning, Washington, DC, pp. 912–919 (2003)
29. Zhu, X., Kandola, J., Ghahramani, Z., Lafferty, J.: Nonparametric transforms of graph kernels for semi-supervised learning. In: Advances in Neural Information Processing Systems, pp. 1641–1648. MIT Press, Cambridge (2005)
30. Zhu, X.: Semi-supervised learning literature survey. Technical report, University of Wisconsin-Madison (2007)
31. Zhou, Z.-H.: A brief introduction to weakly supervised learning. Natl. Sci. Rev. **5**(1), 44–53 (2018)

LSH-Based Graph Partitioning Algorithm

Weidong Zhang$^{(\boxtimes)}$ and Mingyue Zhang

Peking University, Beijing 100871, People's Republic of China
{zhangwd,mingyuezhang}@pku.edu.cn

Abstract. The well-partitioned graph is capable of accelerating the parallel graph algorithms significantly, but few of them adopt the well partitioning algorithms in large scale graph computing. The high time complexity, which even exceed that of the final algorithms occasionally, is the main factor to prevent their applicabilities. Existing graph partitioning algorithms are mostly based on multilevel k-way scheme or iterative label propagation. Most of these algorithms can yield a high-quality result, but the high time/space complexities limit their applications in big data. In this paper, we propose the locality-sensitive hashing (LSH) based graph partitioning algorithm whose time/space complexity is O(n), n is the number of vertices in graph. For all kinds of hyperscale graphs, it works at the speed of random partitioning method approximately. Compared with the latest mainstream graph partitioning algorithms, the new algorithm owns a simple processing pipeline and avoids irregular memory access generated by graph traversals. The experimental result show that the new algorithm achieves 10x faster than Metis and 2x faster than label propagation algorithm at the cost of reasonable precision loss.

Keywords: Graph partitioning algorithm · Locality-sensitive hashing

1 Introduction

In the era of big data and AI, parallelization plays a key role in making algorithms work on big data. The parallelization of algorithms generally requires pre-partitioning on input data according to parallel granularity. The quality of partitioning directly impacts the performance of parallelization. The partitioning quality is measured by the dependencies and balances among the data partitions. Under-optimized partitioning algorithms bring in a large number of dependencies, which increase the communication burden among processors, and then increase the time of data preparation for each round of computation. In addition, when the increased communication overhead counteracts the gain from parallel computing, the larger-scale parallelism becomes meaningless.

For graph partitioning, since Berger et al. proposed the classic Recursive Coordinate Bisection (RCB) algorithm [1], many effectual high-performance algorithms have been proposed, such as Metis [2] which is based on multilevel k-way methods [3,4] and label propagation algorithm (LPA) [5,6]. The time

© Springer Nature Singapore Pte Ltd. 2018
Z.-H. Zhou et al. (Eds.): ICAI 2018, CCIS 888, pp. 55–68, 2018.
https://doi.org/10.1007/978-981-13-2122-1_5

complexities of the multilevel k-way method and label propagation are O(n + m + klog(k)) [3,4] and $O(n + m) + O(nlog_2(n))$ [5] respectively, where m, n, k are the numbers of edges, vertices and partitions. For small-scale datasets, the above two methods both are good options, but when the scale of graph increases to ten million vertices, the evaluation results show that Metis can not work out the final results, and the algorithm based on the label propagation takes more than one second to get the final result. The author of Metis proposed an improved parallel version of Parmetis [7] for large-scale graphs with a higher complexity of $O((n + m) \times log(k))$. Although the label propagation algorithm could handle the graphs with millions of vertices, it requires several times of graph traversals which generate a large number of irregular memory access. Irregular memory access lowers the ratio of computation to memory access, then makes memory access become the performance bottleneck. The released results from the benchmark of pmbw [8] show that the sequential memory access is 30x faster than irregular memory access on various platforms. Although most algorithms claim to be eligible for large-scale graph, few of them is adopted. The reason is that the executing time of graph partitioning algorithm tends to exceed that of the final algorithm, which makes the partitioning meaningless. This paper proposes a LSH-based graph partitioning algorithm, whose time complexity is close to the random graph partitioning algorithm which is widely used in real big data applications. In addition, the new algorithm owns an outstanding scalability that enables it to scale in large-scale graph partitioning problems.

LSH, widely used in mass data retrieval, maps similar data into adjacent locations in hash table. On the other hand, we find that the adjacent vertices in graph could be presented as similar items. So we first utilize the LSH to map the adjacent vertices in graph to adjacent locations in hash space. Then, by partitioning hash table, we get a rough partitioning result of original graph. During the process, the vertex presentation actually performs a same function as a rise-dimensional operation to better preserve the adjacency information in graph; next, a dimensionality reduction operation is performed on the hash space to facilitate partitioning high-dimensional hash space. The LSH-based partitioning algorithm has obvious advantages, which is its linear time complexity, and at the same time, it avoids the irregular memory access introduced by graph traversals. Contributions:

1. This paper proposes a fresh idea to design graph partitioning algorithms, in addition, implements a new graph partitioning algorithm based on LSH.
2. The new graph partitioning algorithm has the following advantages:
   - Low time complexity close to the random graph partitioning algorithm;
   - Avoids irregular memory access due to operations such as graph traversals;
   - Good scalability;
   - Fine partitioning result close to the mainstream partitioning algorithms on large scale graphs.

2   Related Work

Based on different theories and application scenarios, researchers have proposed a number of different graph partitioning algorithms, which roughly fall into three categories: (1) Global Methods: such as the algorithms based on width-first search and combinatorial optimization, these methods are mostly used to solve the small-scale graph partitioning problems [9,10]; (2) The partitioning algorithms based on computational geometry and linear algebra theory, such as RCS, spectrum-based graphing algorithms [11] and multilevel algorithms; (3) The partitioning algorithms based on genetic algorithm [12]. In this section, we will briefly introduce and contrast them with the LSH-based partitioning algorithm.

2.1   Approximately Graph Partitioning Algorithm

Approximate partitioning algorithm only claims to return a approximately accurate result. And at the cost of a small loss of accuracy, it promises an obvious advantage in complexity. The LSH-based graph partitioning algorithm in this paper belongs to this category. Compared with other approximate algorithms, the LSH-based algorithm further lowers the complexity, and avoids the irregular memory access caused by graph traversals. The common approximate graph partitioning algorithms are:

Recursive Coordinate Bisection (RCB). As shown in Fig. 1a, RCB is based on geometric partitioning. The geometric coordinates are first divided into two balanced parts. In each partition, bisection continues recursively until the required number of balanced partitions are obtained.

The Spectral-Based Method. As shown in Fig. 1b, the spectrum partitioning algorithm is based on the intuition that the second lowest vibrational mode of a vibrating string naturally divides the string into two halves [11]. Applying this intuition to the eigenvector of a graph, a new partitioning algorithm is obtained.

The spectral-based graph partitioning algorithm, solving and applying the eigenvectors, essentially leverages the idea of dimensionality reduction on the original graph. Therefore, we classify the spectral-based partitioning algorithm into approximate algorithm. The KMeans algorithm could be adopted to solve eigenvectors, the complexity is O(nkt), n, k, t are the numbers of objects, clusters and iterations respectively. Even if the improved Lanczos algorithm [9] is adopted to solve the eigenvectors, the executing time is still unacceptable for the extremely high dimensional laplacian matrix [10].

The Network Flow Based Graph Partitioning Algorithm [13]. These algorithms cannot work on hypergraphs, and they likely lead to unbalanced partitions. Repeated minimum cuts computation also produces large time overhead.

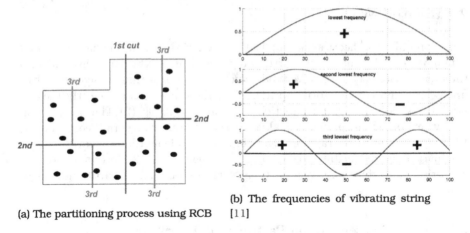

(a) The partitioning process using RCB

(b) The frequencies of vibrating string [11]

Fig. 1. Approximately graph partitioning algorithms

2.2   High Quality Graph Partitioning Algorithm

Kernighan-Lin Algorithm (KLA) [14]: KLA is a simple heuristic optimization algorithm. It refines the partitioning result by the greedy algorithm. KLA firstly divide the graph $G = (V, E, W)$ into two equal parts: $|A| = |B|$. Let $w(e) = w(i,j)$ be the weight of edge $e(i,j)$, and zero if there is no edge. The goal is to find the same number of subsets X in A and Y in B, swapping X and Y to reduce the overhead of AB-community edge connections.

Metis: It is a toolkit consisting of several graph partitioning programs. The implementation of these programs is based on multilevel algorithm. Metis' multilevel algorithm consists of three stages, each of which has several different implementations [15].

Label Propagation Algorithm (LPA): LPA was originally proposed for graph clustering. A simple LP first assigns an unique label id to each vertex, then updates the label as the one that is prevalent in its neighborhood iteratively until no label changes. Vertices sharing the same label are grouped into the same partition [16].

2.3   Comparison with Other Partitioning Algorithms

To provide an overview, we compare it with some more algorithms in four aspects: time/space complexities, quality of result and pattern of memory access. The result is shown as in Table 1.

Table 1. Comparison between common graph partitioning algorithms and LSH-based graph partitioning algorithm in four aspects: time/space complexities; quality of partitioning result; the pattern of memory access.

Algorithm	Time	Space	Result	Mem access
Random partition	$O(n)$	$O(m)$	Poor	Regular
Kernighan-Lin	$O(n^3)$ [14]	$O(m)$	Good	Irregular
Spectral	$O(n^3)$	$O(n^2)$	Good	Irregular
LSH-based	$O(n)$	$O(m)$	Good	Regular
Metis	$O(m \log(n/n'))$ [2]	$O(m)$	Excellent	irregular
LPA	$O(n + m) + O(n log_2(n))$	$O(m)$	Excellent	Irregular

3   Models and Algorithms

The motivation of this work is to find a way of partitioning graph with extremely low complexity without compromising observable accuracy loss. Our solution adopts the LSH. It first maps the adjacent vertices in graph into adjacent locations in hash space, then partitions the hash space to get the result.

Figure 2 shows a tiny sample graph, in which each vertex is presented as an item in the second column of Table 2. We find that the presentations of adjacent vertices in graph show certain similarity. The similarity inspires us to utilize the LSH to partition graphs. By combining the LSH with the similar presentations of adjacent vertices, it works out a rough partitioning result. The entire process involves three key steps:

– Choose an appropriate vertex presentation which not only expresses the similarity of adjacent vertices, but also delivers sufficient distinctiveness of non-adjacent vertices.
– Select an appropriate locality-sensitive hash function cluster.
– Design a suitable hash space partitioning method.

The following part in this section describes in detail the methods we have taken to settle the above three steps.

3.1   Vertex Numbering and Presentation

As shown in Fig. 2 and Table 2, each vertex is presented by all its neighbor vertices. In this way, each vertex gets an unique simple presentation in most of graphs. Whereas, its expressiveness is limited, it can not express the positional relationships among the vertices sharing no common vertices. This subsection explores some more expression approaches. The goal is to find a presentation carrying sufficient information to deliver:

– Adequate similarity among adjacent vertices.
– Decreasing similarities as the vertices' distance increases gradually.

Here, we use the Jaccard distance to measure the similarity of two vertices.

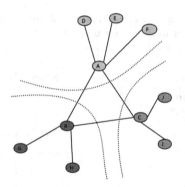

Fig. 2. Demo graph

Table 2. Similarities between adjacent vertices [17]. Each vertex is presented by all the neighbors.

VID	Vertex presentation	Jaccard Similarity									
		A	B	C	D	E	F	G	H	I	J
A	$\langle A,B,C,D,E,F \rangle$	1	.37	.37	.33	33	.33	.14	.14	.14	.14
B	$\langle A,B,C,G,H \rangle$	.37	1	.42	.16	.16	.16	.4	.4	.16	.16
C	$\langle A,B,C,I,J \rangle$	.37	.42	1	.16	.16	.16	.16	.16	.4	.4
D	$\langle A,D \rangle$	.33	.16	.16	1	.33	.33	0	0	0	0
E	$\langle A,E \rangle$	.33	.16	.16	.33	1	.33	0	0	0	0
F	$\langle A,F \rangle$	.14	.16	.16	.33	.33	1	0	0	0	0
G	$\langle B,G \rangle$	.14	.4	.16	0	0	0	1	.33	0	0
H	$\langle B,H \rangle$	.14	.4	.16	0	0	0	.33	1	0	0
I	$\langle C,I \rangle$	.14	.16	.4	0	0	0	0	0	1	.33
J	$\langle C,J \rangle$	.14	.16	.4	0	0	0	0	0	.33	1

Vertex Numbering. The integer is commonly used to uniquely indicate a vertex in graph. The advantage of this scheme is that it is exquisite and friendly to storage and computing resources. The disadvantage, from the view of the informatics, is that the minimum elements "digit: 0–9" are too few to make the vertices' presentations show sufficient similarity and distinctiveness. Inspired by text retrieval, we attempt to replace digits with letters.

- **Use a letter to encode a number:** This scheme uses {a, b, c, d, e, f, g, h, i, j} to encode {0, 1, 2, 3, 4, 5, 6, 7, 8, 9}. For example, a vertex whose number is 199 will be converted to "bjj". Using the presentation method in Table 2, a vertex will be encoded as a text document.
- **Use multiple distinct letters or digits to encode a number:** In this scheme, {0, 1, 2, 3, 4, 5, 6, 7, 8, 9} will be encoded with more than one distinct

letters, like {ab, cd, ef, gh, ij, kl, mn, op, qr, st}, {ab1, cd2, ef3, gh4, ij5, kl6, mn7, op8, qr9, st0}, etc.

Vertex Presentation. In order to bring in more positional information and increase the similarity as the vertices' distance decrease, we design several strategies to present the vertices.

- **Tree-like presentation:** This scheme expresses each vertex as a set of vertices, which are located in the scope of a tree structure with the current vertex as the root node. The location information of vertex can be adjusted by the depth of the tree.
- **Star-like presentation:** This scheme expresses each vertex as a set of vertices, which are located in the scope of a star structure centered on the current vertex. The location information of vertex can be adjusted by the radius of the star.

The depth of tree and the radius of star should both be proportional to the scale and shape of target graphs.

Other Vertex Presentations. Encoding vertices using tree and star structures leads to an exponential increase to presentation's length as depth increases. A compromise is to random select parts of the paths or branches in its structure so that a reasonable length of vertex representation is ensured, and at the same time the similarity within a certain distance is guaranteed.

3.2 Mapping Adjacent Vertices to Hash Space

Nearest neighbor search is a routine task in massive high-dimensional data processing. Different from low-dimensional data, it is time consuming if the conventional distance-based sequential lookup is used for high-dimensional data. LSH is able to map similar data to adjacent buckets of hash table by high probability, and map the data with low similarity to adjacent buckets by extremely low probability. Using the characteristic of LSH, the original linear search with complexity $O(n)$ can be substituted by hashing mapping in constant-order time complexity.

The principle of LSH is not complicated. The key is to find a cluster of hash functions for corresponding application scenario. Such a family of hash functions $H = \{h:S \rightarrow U\}$ is (r_1, r_2, p_1, p_2) sensitive, if for any function h in H satisfies the following two conditions:

- If $d(O_1, O_2) < r_1, Pr[h(O_1) = h(O_2)] \geq p_1$
- If $d(O_1, O_2) > r_2, Pr[h(O_1) = h(O_2)] \geq p_2$

where $O_1, O_2 \in S$ present two data objects with multi-dimensional attributes, $d(O_1, O_2)$ is the distance between two objects. The two conditions actually state a constraint that when sufficiently similar, the probability of mapping to the

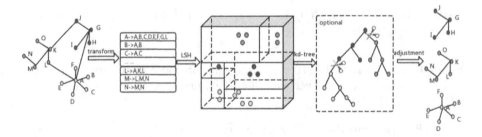

Fig. 3. Pipeline of LSH-based partitioning algorithm

same bucket is large enough; when sufficiently dissimilar, the probability of mapping to the same bucket is sufficiently small. The definition of distance function depends on the actual application situation.

To ensure reasonable recall rate, LSH uses a set of hash functions to map every data item into multiple hash tables. As long as the similar data items are mapped into at least one same bucket or adjacent buckets in all the hash tables, the recall rate is guaranteed. To recall data with reasonably lower similarity, LSH needs more hash tables, each of which contains the labels of all the vertices in graph. More hash tables bring in higher space complexity. To solve the problem of high space complexity, LSH Forest [18] and Multi-prob LSH [19] leverage the dictionary tree and nearest query.

As is shown in Fig. 3, we first express the graph vertices as text documents, then map them to the hash space by LSH. The hash space of LSH can be considered as a high-dimensional vector space. In spite of the fact that items with low similarities cannot be mapped to the same hash value, partial dimensions of their hash values still have certain similarities. This is the reason why LSH Forest and Multi-prob LSH work. So, after presenting and mapping, the adjacent vertices in graph are mapped to adjacent buckets in high-dimensional space.

3.3   Indexing Hash Value by kd-tree

Through the mapping of LSH, the adjacent vertices have been mapped to the adjacent buckets in hash table. Then, we could partition the graph by partitioning the hash space. Yet, the LSH value space is a high-dimensional vector space, it is hard to partition the high-dimensional space according to the number of data items. So we further index the hash values with tree-like structure index. The tree-like index arranges the adjacent hash values in the same subtrees. After inorder traversal, the adjacent points on tree, in a sense, are mapped to adjacent positions in linear sequence. By dividing the linear sequence according to the number of data items, a preliminary partitioning result can be obtained. The common high-dimensional tree-like indexes are kd-tree [20], R-tree [21], X-tree [22], etc. We choose the kd-tree to index these high-dimensional hash values for its conciseness.

3.4   Adjusting the Preliminary Result

Only one hash table of LSH is used during creating kd-tree. The elements in the rest of hash tables will be used as the baseline for adjustment in this step. The adjustment process is as follows:

- traverses the rest of LSH hash tables, if the vertices in a bucket are not unique, check the partitions that the vertices belong to.
- if these vertices belong to a same partition, no operation is performed.
- otherwise moves all vertices to the partition which more than half of the vertices belong to.

Repeat above operations until movement converges. The time complexity of the adjustment is $O(n)$, n is the number of vertices.

In the experiment stage, we find that even without indexing by and partitioning on kd-tree, a forthright adjustment on LSH's result can achieve a approximate result to the former. In the following section, the implementation of the LSH-based partitioning algorithm simply consists of vertex presenting, LSH mapping and adjusting.

4   Evaluation and Result Analysis

4.1   Experimental Platform and Datasets

All the evaluations are carried out on an instance of m4.16xlarge type from cn-north-1 of AWS EC2 Service. The instance is equipped with 64×vCPUs (Intel Xeon E5-2686v4), 256 GiB memory. The datasets come from the SNAP Datasets [23]. Five graphs at different scales are selected, the basic statistical figures are demonstrated in Table 3.

Three mainstream partitioning algorithms are adopted to compare and evaluate the new algorithm's performance, which include Metis, label propagation based algorithm [24] and random partitioning algorithm.

Table 3. Basic statistics of dataset.

Dataset	#vertex	#edges	#AvgNgb	Size
new_sites_edges	27,917	206,259	7.39	2.5M
gemsec-deezer	143,884	846,915	5.88	4.59M
wiki-talk	1,140,149	1,654,796	1.45	21M
twitter_rv	61,578,414	1,468,365,182	23.85	25G

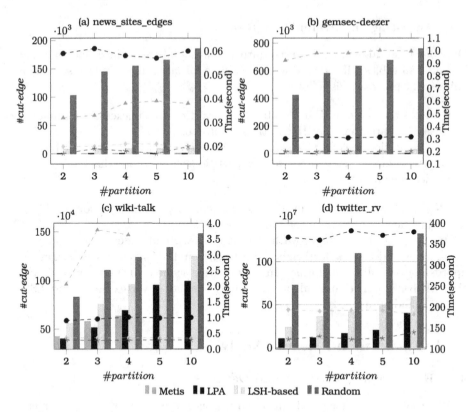

Fig. 4. Partitioning result. The histograms demonstrate the numbers of cut edges between partitions, the wave lines denote the partitioning time.

4.2 Experiment Results and Analysis

We compare the algorithms mainly in three aspects including: cut edge, balance and partitioning time. The balance is defined as the ratio of the largest partition's size to the partitions' average size. In terms of time, we list I/O time and partitioning time. The numbers of partitions include 2, 3, 4, 5, 10.

According to the results listed in Table 4 and Fig. 4, we roughly classify the results to two groups:

(1.) The performance on the graphs with less than one million vertices, shown as in Fig. 4(a,b)**:**

- cut edge: Compared with random partitioning algorithm, the new algorithm slashes the cut edges by 95%, but still generates 7x–15x cut edges than Metis and LPA.
- partitioning time: The new algorithm run 1.5x–3x faster than Metis and LPA, and almost as fast as random partitioning algorithm.
- balance: The new algorithm performs best among these algorithms in keeping balance.

Table 4. The partitioning results. '-' denotes that the algorithm failed working on the graphs with the specified partitions.

Algorithm	#partitions	#cut-edges	Cut-edge ratio	Balance	IO time	Partition time (second)
	news_sites_edges					
Metis	2	551	0.0026	1.00	0.019	0.032
	3	649	0.0031	1.00	0.020	0.033
	4	841	0.0041	1.01	0.019	0.038
	5	1044	0.0051	1.00	0.022	0.039
	10	1259	0.0061	1.006	0.020	0.038
LSH partition	2	8418	0.0408	1.00	0.021	0.020
	3	9055	0.0439	1.00	0.022	0.020
	4	9106	0.0441	1.00	0.022	0.021
	5	9131	0.0442	1.00	0.022	0.021
	10	9893	0.0479	1.00	0.022	0.019
Random partition	2	102923	0.4990	1.00	0.019	0.017
	3	144588	0.701	1.00	0.020	0.019
	4	154694	0.75	1.00	0.020	0.018
	5	165626	0.803	1.00	0.021	0.017
	10	185633	0.90	1.00	0.019	0.020
Label propagation	2	796	0.0039	1.001	0.020	0.059
	3	989	0.0048	1.003	0.019	0.061
	4	1051	0.0051	1.001	0.019	0.058
	5	1189	0.0058	1.002	0.020	0.057
	10	1217	0.0059	1.001	0.021	0.060
	gemsec-deezer					
Metis	2	1013	0.0012	1.00	0.191	0.923
	3	2615	0.0031	1.00	0.192	0.981
	4	2829	0.0033	1.01	0.192	0.981
	5	3099	0.0037	1.00	0.189	1.003
	10	4123	0.0049	1.00	0.179	0.997
LSH partition	2	20893	0.024	1.00	0.212	0.198
	3	21001	0.025	1.00	0.198	0.186
	4	21289	0.025	1.00	0.211	0.197
	5	23870	0.028	1.00	0.215	0.198
	10	30186	0.036	1.00	0.218	0.201
Random partition	2	423457	0.50	1.00	0.199	0.201
	3	582677	0.688	1.00	0.209	0.199
	4	634339	0.749	1.00	0.200	0.202
	5	677532	0.80	1.00	0.210	0.197
	10	761376	0.899	1.00	0.200	0.204
Label propagation	2	1196	0.0014	1.001	0.209	0.301
	3	2908	0.0034	1.02	0.213	0.317
	4	3810	0.0045	1.004	0.214	0.308
	5	4001	0.0047	1.001	0.215	0.314
	10	5201	0.0062	1.006	0.216	0.316

(continued)

Table 4. (*continued*)

Algorithm	#partitions	#cut-edges	Cut-edge ratio	Balance	IO time	Partition time (second)
	wiki-talk					
Metis	2	424357	0.256	1.03	0.306	2.064
	3	576704	0.348	1.03	0.300	3.779
	4	630843	0.381	1.03	0.294	3.622
	5	–	–	–	–	–
	10	–	–	–	–	–
LSH partition	2	557514	0.336	1.00	0.422	0.367
	3	752884	0.454	1.00	0.414	0.360
	4	954840	0.577	1.00	0.415	0.359
	5	1098297	0.663	1.00	0.416	0.354
	10	1245994	0.752	1.00	0.416	0.348
Random partition	2	827398	0.500	1.00	0.410	0.289
	3	1103749	0.667	1.00	0.402	0.276
	4	1234477	0.746	1.00	0.419	0.280
	5	1338729	0.809	1.00	0.408	0.291
	10	1481042	0.895	1.00	0.412	0.289
Label propagation	2	401381	0.243	1.003	0.510	0.906
	3	513465	0.311	1.00	0.500	0.953
	4	689156	0.417	1.003	0.498	1.010
	5	951874	0.575	1.00	0.501	0.984
	10	993456	0.600	1.001	0.499	1.003
	twitter_rv					
Metis	2,3,4,5,10	–	–	–	–	–
LSH partition	2	237366143	0.161	1.04	258.65	193.262
	3	362323534	0.246	1.00	257.580	189.888
	4	416724538	0.283	1.002	258.955	192.118
	5	488875795	0.332	1.09	264.413	192.247
	10	595942924	0.405	1.08	257.231	181.907
Random partition	2	728309130	0.496	1.00	256.364	123.156
	3	976462846	0.665	1.00	255.301	130.155
	4	1095400428	0.746	1.00	261.659	123.069
	5	1179097241	0.803	1.00	259.833	125.734
	10	1327402124	0.904	1.00	250.183	139.701
Label propagation	2	105678319	0.072	1.001	261.843	366.245
	3	113897462	0.078	1.02	259.270	359.135
	4	164538716	0.112	1.001	264.161	381.975
	5	201698756	0.137	1.00	267.632	370.927
	10	399345618	0.272	1.00	267.625	379.271

- partition: As the number of partitions increases, the number of cut edges also tends to increase for all the algorithms. The new algorithm presents the slowest increment speed among them.

(2.) The performance on the graphs with more than one million vertices, shown as in Fig. 4(c,d):

- cut edge: The new algorithm could reduce the cut edges by 83% from the result of random partitioning algorithm, and generates about 33% more cute edges than LPA.

– partitioning time: The new algorithm works 2.8x–10x faster than LPA, and 30% slower than random partitioning algorithm.
– balance: Same as in (1.).
– partition: Same as in (1.).

In this group, Metis fail to return some parts of the final results for its high complexity.

5 Conclusion

The LSH-based graph partitioning algorithm proposed in this paper owns a very low time complexity of $O(n)$, the partitioning speed is close to that of random partitioning algorithm. For various large scale graphs, the quality of partitioning result is close to that of high-complexity algorithms. At the same time, we find that the strategy of adjustment plays a key role to the quality of the final result. In the future work, more adjustment strategies for different types of graphs will be tested and evaluated to improve the partitioning result.

References

1. Berger, M.J., Bokhari, S.H.: A partitioning strategy for nonuniform problems on multiprocessors. IEEE Computer Society (1987)
2. Karypis, G., Kumar, V.: Metis: a software package for partitioning unstructured graphs. In: International Cryogenics Monograph, pp. 121–124 (1998)
3. Sanchis, L.A.: Multiple-way network partitioning. IEEE Trans. Comput. **38**(1), 62–81 (1989)
4. Complexity of pmetis and kmetis algorithms, 27 March 2018. http://glaros.dtc. umn.edu/gkhome/node/419
5. Zhang, X.K., Ren, J., Song, C., Jia, J., Zhang, Q.: Label propagation algorithm for community detection based on node importance and label influence. Phys. Lett. A **381** (2017)
6. Backstrom, L., Backstrom, L.: Balanced label propagation for partitioning massive graphs. In: ACM International Conference on Web Search and Data Mining, pp. 507–516 (2013)
7. Karypis, G., Schloegel, K., Kumar, V.: Parmetis: parallel graph partitioning and sparse matrix ordering library, Version, Department of Computer Science, University of Minnesota (2003)
8. pmbw - parallel memory bandwidth results, 27 March 2018. https://panthema. net/2013/pmbw/results.html
9. Lanczos, C.: An iteration method for the solution of the eigenvalue problem of linear differential and integral operators. J. Res. Natl. Bureau Stand. **45**(45), 255–282 (1950)
10. Buluc, A., Meyerhenke, H., Safro, I., Sanders, P., Schulz, C.: Recent advances in graph partitioning, vol. 77, no. 1, pp. 207–220 (2013)
11. Kabelíková, P.: Graph partitioning using spectral methods (2006)
12. Jin, K., Hwang, I., Kim, Y.H., Moon, B.R.: Genetic approaches for graph partitioning: a survey. In: Proceedings of Genetic and Evolutionary Computation Conference, GECCO 2011, Dublin, Ireland, July 2011, pp. 473–480 (2011)

13. Network flow based partitioning, 27 March 2018. http://users.ece.utexas.edu/~dpan/EE382V_PDA/notes/lecture5_partition-networkflow.pdf
14. Kernighan, B.W., Lin, S.: An efficient heuristic procedure for partitioning graphs. Bell Syst. Tech. J. **49**(2), 291–307 (1970)
15. Metis-wikipedia, 27 March 2018. https://en.wikipedia.org/wiki/METIS
16. Wang, L., Xiao, Y., Shao, B., Wang, H.: How to partition a billion-node graph (2014)
17. Zhang, W., He, B., Chen, Y., Zhang, Q.: GMR: graph-compatible mapreduce programming model. Multimedia Tools Appl. **1**, 1–19 (2017)
18. Bawa, M., Condie, T., Ganesan, P.: LSH forest: self-tuning indexes for similarity search. In: International Conference on World Wide Web, pp. 651–660 (2005)
19. Lv, Q., Josephson, W., Wang, Z., Charikar, M., Li, K.: Multi-probe LSH: efficient indexing for high-dimensional similarity search. In: International Conference on Very Large Data Bases, pp. 950–961 (2007)
20. Bentley, J.L.: Multidimensional binary search trees used for associative searching. Commun. ACM **18**(9), 509–517 (1975)
21. Guttman, A.: R-trees: a dynamic index structure for spatial searching. In: ACM SIGMOD International Conference on Management of Data, pp. 47–57 (1984)
22. Berchtold, S., Keim, D.A., Kriegel, H.P.: The X-tree: an index structure for high-dimensional data. In: Proceedings of VLDB, September 1996, Mumbai, India, pp. 28–39 (1996)
23. Leskovec, J., Krevl, A.: SNAP Datasets: Stanford large network dataset collection, June 2014. http://snap.stanford.edu/data
24. Raghavan, U.N., Albert, R., Kumara, S.: Near linear time algorithm to detect community structures in large-scale networks. Phys. Rev. E Stat. Nonlinear Soft Matter Phys. **76**(2), 036106 (2007)

Graph Regularized Discriminative Joint Concept Factorization for Data Representation

Xianzhong Long[1,2,3(✉)] and Cheng Cheng[1]

[1] School of Computer Science and Technology,
Nanjing University of Posts and Telecommunications, Nanjing 210023, China
lxz@njupt.edu.cn, chengchengchn@163.com
[2] College of Computer Science and Technology,
Nanjing University of Aeronautics and Astronautics, Nanjing 211106, China
[3] Jiangsu Key Laboratory of Big Data Security and Intelligent Processing,
Nanjing 210023, China

Abstract. In the computer vision and pattern recognition fields, how to effectively represent the data is very important. A good data representation scheme can improve subsequent classification or clustering tasks. In this paper, a graph regularized discriminative joint concept factorization (GDJCF) method is proposed for data representation. In the GDJCF, we make use of three aspects of information, including multi-features joint learning, local manifold geometry structure and sample label. The learned representations are expected to have discriminative ability and better describe the raw data. We provide the corresponding multiplicative update solutions for the optimization framework, together with the convergence verification. Clustering results on five image data sets show that the GDJCF outperforms the state-of-the-art algorithms in terms of accuracy and normalized mutual information.

Keywords: Joint concept factorization · Multi-features learning
Graph regularized · Clustering

1 Introduction

In the machine learning, some effective data representation schemes were presented, such as sparse coding (SC) [24], low rank representation (LRR) [15], non-negative matrix factorization (NMF) [23], concept factorization (CF) [14] and so on. In the SC [24], one test sample can be possibly represented as a linear combination of just those training samples from the same class and this representation is naturally sparse. The sparse representation can be obtained by solving a L_1 minimization problem. Different from sparse coding, LRR aims at finding the lowest-rank representation and the computational procedure of LRR is to solve a L_* norm (nuclear norm) regularized optimization problem [15]. NMF

Z.-H. Zhou et al. (Eds.): ICAI 2018, CCIS 888, pp. 69–83, 2018.
https://doi.org/10.1007/978-981-13-2122-1_6

incorporates the nonnegativity constraints and thus obtains the parts-based representation. The existing NMF algorithms are divided into four categories in the [23] and local optimal solutions are achieved for these NMF variants. CF can be considered as a special case of NMF and it decomposes a nonnegative matrix into the form of multiplication of three nonnegative matrix [14]. In the CF, there are two coefficient matrices which are used to represent the dictionary and the original sample respectively.

The data representation methods mentioned above can be seen as a matrix factorization with different constraints and have respective advantage or disadvantage. For example, the SC and LRR have better representation ability, but the process of solving the corresponding optimization problems are very time consuming. Although solving the NMF and CF problem is very fast, they can only obtain the local optimal solutions. We should choose one of them according to the actual application. In this paper, we focus on CF and propose an effective data representation strategy based on multi-features, manifold learning and label information.

The parts-based representation property of NMF is consistent with the psychological and physical evidence in the human brain. NMF has been applied in many real world problems such as face recognition [27], image clustering [16], recommender systems [19], speech emotion recognition [22]. In spite of the fact that NMF has been widely used, there are still some disadvantages. The major limitation of NMF is that it is unclear how to effectively perform NMF in the transformed data space, e.g. reproducing kernel Hilbert space (RKHS) [5]. To address the limitations of NMF while inheriting all its strengths, CF is proposed for data clustering [26]. It can be seen as a variant of nonnegative matrix factorization. In CF, each concept is modeled as a nonnegative linear combination of the data points, and each data point is represented as a linear combination of the concepts. The major advantage of CF over NMF is that it can be performed on any data representations, either in the original space or RKHS.

CF is an unsupervised learning method, which has been widely employed and some variants based on it have been proposed successively. For the sake of enhancing the discriminative abilities of the image representations, class-driven concept factorization (CDCF) technology was proposed in [13]. The class-driven constraint forces the representations of data points to be more similar within the same class while different between classes. Another semi-supervised concept factorization (SSCF) algorithm was given in [18]. SSCF incorporates the pairwise constraints into CF as the reward and penalty terms, which can guarantee that the data points belonging to a cluster in the original space are still in the same cluster in the transformed space. In order to maintain geometrical neighborhood structure of the data, the CF with adaptive neighbors (CFAN) was presented in [20]. CFAN performs dimensionality reduction and finds the neighbor graph weights matrix simultaneously. Graph regularized multilayer concept factorization (GMCF) was proposed in [14]. GMCF is a multi-stage procedure, which decomposes the observation matrix iteratively in a number of layers. For utilizing the local manifold regularization, graph-based discriminative concept factorization (GDCF) was put forward in [12]. GDCF not only encodes the local geometrical structure of the data space by constructing K-nearest graph, but also takes into account the available label information.

Recently, joint nonnegative matrix factorization (JNMF) has been applied in many fields such as multi-view clustering [17], microbiome data analysis [8], latent information discovery [7], exemplar-based voice conversion [25] and topic modeling [9]. However, as far as we know, joint concept factorization has not been proposed in the existing literatures. Meanwhile, label information and graph Laplacian regularized term are also very important [3,21]. Inspired by these, in this paper, we present a joint concept factorization method based on graph regularized discriminative information.

The remainder of this paper is organized as follows: Sect. 2 introduces the basic idea of existing clustering algorithms. Our graph regularized discriminative joint concept factorization scheme is presented in Sect. 3. In Sect. 4, the comparative results of clustering on five widely used image data sets are reported. Finally, conclusions are made in Sect. 5.

2   Related Work

Let $\mathbf{X}$ be a data matrix of n m-dimensional samples $\mathbf{x}_1, \mathbf{x}_2, \cdots, \mathbf{x}_n$, i.e., $\mathbf{X} = [\mathbf{x}_1, \mathbf{x}_2, \cdots, \mathbf{x}_n] \in \mathbb{R}^{m \times n}$. Each column of $\mathbf{X}$ represents an image with m dimensions. $\mathbf{W} \in \mathbb{R}^{m \times r}$, $\mathbf{U} \in \mathbb{R}^{n \times r}$, $\mathbf{V} \in \mathbb{R}^{n \times r}$ and $\mathbf{H} \in \mathbb{R}^{r \times n}$ ($r \ll \min(m, n)$). This section reviews the k-means, non-negative matrix factorization, graph regularized non-negative matrix factorization, concept factorization and locally consistent concept factorization algorithms.

2.1   K-Means

K-Means clustering [6] aims to partition the n observations into $r(r \leq n)$ sets $\mathbf{S} = \{S_1, S_2, \ldots, S_r\}$ so as to minimize the within-cluster sum of squares. Formally, the objective is to find:

$$\min_{\mathbf{S}} \sum_{i=1}^{r} \sum_{\mathbf{x} \in S_i} \|\mathbf{x} - \mu_{\mathbf{i}}\|_2^2 \tag{1}$$

Where μ_i is the mean of points in S_i. After K-Means clustering, each observation is assigned to a unique center.

2.2   Non-negative Matrix Factorization

NMF [11] decomposes a matrix $\mathbf{X}$ into a product of two non-negative matrices $\mathbf{W}$ and $\mathbf{H}$, i.e., $\mathbf{X} \approx \mathbf{W}\mathbf{H}$. Some algorithms for NMF have been proposed in [2,10], such as multiplicative update rules, gradient descent and alternating least squares.

Multiplicative update rules were firstly considered in [10]. The criterion function of NMF is based on minimizing the Euclidean distance between $\mathbf{X}$ and $\mathbf{WH}$. The corresponding optimization problem is as follows:

$$\min_{\mathbf{W},\mathbf{H}} \|\mathbf{X} - \mathbf{WH}\|_F^2$$
$$\text{s.t.} \quad \mathbf{W} \geq 0, \mathbf{H} \geq 0 \tag{2}$$

Where $\|\cdot\|_F$ denotes the matrix Frobenius norm, $\mathbf{X}$, $\mathbf{W}$ and $\mathbf{H}$ is called sample matrix, basis matrix and coefficient matrix respectively. In the machine learning field, $\mathbf{W}$ can be used as a projected matrix to reduce the dimension of data, and $\mathbf{H}$ is used to clustering.

The well-known multiplicative update rules for solving (2) are as follows [10,11]:

$$\mathbf{W}_{iq} \longleftarrow \mathbf{W}_{iq} \frac{(\mathbf{XH}^T)_{iq}}{(\mathbf{WHH}^T)_{iq}}$$
$$\mathbf{H}_{qj} \longleftarrow \mathbf{H}_{qj} \frac{(\mathbf{W}^T\mathbf{X})_{qj}}{(\mathbf{W}^T\mathbf{WH})_{qj}} \tag{3}$$

The convergence of these multiplicative update rules have been proved in [10].

2.3 Graph Regularized Non-negative Matrix Factorization

Recently, GNMF [3] encodes the data geometric structure in an nearest neighbor graph. It solved the following optimization problem:

$$\min_{\mathbf{W},\mathbf{H}} \|\mathbf{X} - \mathbf{WH}\|_F^2 + \lambda \text{Tr}(\mathbf{HLH}^T)$$
$$\text{s.t.} \quad \mathbf{W} \geq 0, \mathbf{H} \geq 0 \tag{4}$$

Where $\mathbf{L}$ is the graph Laplacian matrix [1] and $\mathbf{L} = \mathbf{B} - \mathbf{C}$. Here, $\mathbf{C}$ is a weight matrix, and $\mathbf{B}$ is a diagonal matrix whose entries are column or row sum of $\mathbf{C}$. The Tr is the trace of matrix, i.e., the sum of matrix diagonal entries.

The corresponding multiplicative update rules for solving (4) are as follows:

$$\mathbf{W}_{iq} \longleftarrow \mathbf{W}_{iq} \frac{(\mathbf{XH}^T)_{iq}}{(\mathbf{WHH}^T)_{iq}}$$
$$\mathbf{H}_{qj} \longleftarrow \mathbf{H}_{qj} \frac{(\mathbf{W}^T\mathbf{X}+\lambda\mathbf{HC})_{qj}}{(\mathbf{W}^T\mathbf{WH}+\lambda\mathbf{HB})_{qj}} \tag{5}$$

2.4 Concept Factorization

CF [26] factorizes a non-negative matrix $\mathbf{X} \in \mathbb{R}^{m \times n}$ into a product of three non-negative matrices, i.e., $\mathbf{X} \approx \mathbf{XUH}$.

$$\min_{\mathbf{U},\mathbf{H}} \|\mathbf{X} - \mathbf{XUH}\|_F^2$$
$$\text{s.t.} \quad \mathbf{U} \geq 0, \mathbf{H} \geq 0 \tag{6}$$

Where each column of the $\mathbf{XU} \in \mathbb{R}^{m \times r}$ is called a concept and each concept is modeled as a nonnegative linear combination of the data points (each column of $\mathbf{X}$). Similarly, each data point is represented as a linear combination of the concepts.

The multiplicative update rules for solving (6) are as follows:

$$\mathbf{U}_{jk} \longleftarrow \mathbf{U}_{jk} \frac{(\mathbf{X}^T\mathbf{X}\mathbf{H}^T)_{jk}}{(\mathbf{X}^T\mathbf{X}\mathbf{U}\mathbf{H}\mathbf{H}^T)_{jk}}$$
$$\mathbf{H}_{kj} \longleftarrow \mathbf{H}_{kj} \frac{(\mathbf{U}^T\mathbf{X}^T\mathbf{X})_{kj}}{(\mathbf{U}^T\mathbf{X}^T\mathbf{X}\mathbf{U}\mathbf{H})_{kj}} \quad (7)$$

Before iteration, we should calculate $\mathbf{K} = \mathbf{X}^T\mathbf{X}$ in advance and $\mathbf{K}$ can be considered as a specific kernel function.

2.5   Locally Consistent Concept Factorization

LCCF was proposed in [5], which solved the following optimization problem:

$$\min_{\mathbf{U},\mathbf{H}} \|\mathbf{X} - \mathbf{X}\mathbf{U}\mathbf{H}\|_F^2 + \lambda\text{Tr}(\mathbf{H}\mathbf{L}\mathbf{H}^T)$$
$$\text{s.t. } \mathbf{U} \geq 0, \mathbf{H} \geq 0 \quad (8)$$

Similar to formula (4), $\mathbf{L}$ is the graph Laplacian matrix and $\mathbf{L} = \mathbf{B} - \mathbf{C}$.
The corresponding multiplicative update rules for solving (8) are as follows:

$$\mathbf{U}_{jk} \longleftarrow \mathbf{U}_{jk} \frac{(\mathbf{X}^T\mathbf{X}\mathbf{H}^T)_{jk}}{(\mathbf{X}^T\mathbf{X}\mathbf{U}\mathbf{H}\mathbf{H}^T)_{jk}}$$
$$\mathbf{H}_{kj} \longleftarrow \mathbf{H}_{kj} \frac{(\mathbf{U}^T\mathbf{X}^T\mathbf{X}+\lambda\mathbf{H}\mathbf{C})_{kj}}{(\mathbf{U}^T\mathbf{X}^T\mathbf{X}\mathbf{U}\mathbf{H}+\lambda\mathbf{H}\mathbf{B})_{kj}} \quad (9)$$

3   Graph Regularized Discriminative Joint Concept Factorization

Inspired by the multi-features joint learning, manifold learning theory and label information, we propose a joint concept factorization scheme based on graph regularized discriminative terms. The definition and update rules of GDJCF are as follows.

3.1   GDJCF Model

The use of complementarity between different features can improve performance. We extract two kinds of features for each image, i.e., $\mathbf{X} \in \mathbb{R}^{m \times n}$ and $\mathbf{Y} \in \mathbb{R}^{d \times n}$. m and d are denoted as dimensionality of features, and n is the number of samples. When we combine $\mathbf{X}$ and $\mathbf{Y}$, we can get feature fusion matrix $\mathbf{Z} = [\mathbf{X}; \mathbf{Y}] \in \mathbb{R}^{p \times n}$, $p = m + d$. Besides, recent studies in manifold learning theory have proved that the local geometric structure can be effectively modeled through a k nearest neighbor graph on a scatter of data points. Consider a graph with n vertices where each vertex corresponds to a data point. For each data point $\mathbf{z}_i$, we find its k nearest neighbors and put edges between $\mathbf{z}_i$ and its neighbors. There are many choices to define a weight matrix $\mathbf{C}$ on the graph, such as '0–1 weighting', 'Heat kernel weighting' and 'Dot-product weighting'.

We construct our weight matrix by using the '0–1 weighting'. The entries of our weight matrix are defined by:

$$\mathbf{C}_{i,j} = \begin{cases} 1, & \text{if } \mathbf{z}_i \in N_k(\mathbf{z}_j) \ or \ \mathbf{z}_j \in N_k(\mathbf{z}_i) \\ 0, & \text{otherwise} \end{cases} \tag{10}$$

Where $N_k(\mathbf{z}_i)$ consists of k NNs of $\mathbf{z}_i$ and they have the same label as $\mathbf{z}_i$. $\mathbf{C} \in \mathbb{R}^{n \times n}$, and a diagonal matrix $\mathbf{B} \in \mathbb{R}^{n \times n}$ is obtained according to weight matrix $\mathbf{C}$. $\mathbf{B}_{ii}$ is the row (or equivalently column, since $\mathbf{C}$ is symmetrical) sum of the weight matrix $\mathbf{C}$, i.e., $\mathbf{B}_{ii} = \sum_{j=1}^{n} \mathbf{C}_{ij}$. $\mathbf{L} = \mathbf{B} - \mathbf{C}$ is called graph Laplacian matrix and $\mathbf{L} \in \mathbb{R}^{n \times n}$.

We also define another class indicator matrix $\mathbf{S} \in \mathbb{R}^{c \times n}$ and give the definition as follows:

$$\mathbf{S}_{i,j} = \begin{cases} 1, & \text{if } t_j = i, \ j = 1, 2, \cdots, n, \ i = 1, 2, \cdots, c \\ 0, & \text{otherwise} \end{cases} \tag{11}$$

Where $t_j \in \{1, 2, \cdots, c\}$ denotes the class label of the jth sample $\mathbf{z}_j$ and c is the total number of classes.

The GDJCF solves the following optimization problem:

$$\begin{aligned} \min_{\mathbf{U}, \mathbf{V}, \mathbf{H}, \mathbf{A}} \ & \alpha \|\mathbf{X} - \mathbf{XUH}\|_F^2 + \beta \|\mathbf{Y} - \mathbf{YVH}\|_F^2 \\ & + \lambda \mathrm{Tr}(\mathbf{HLH}^T) + \gamma \|\mathbf{S} - \mathbf{AH}\|_F^2 \\ \text{s.t.} \ & \mathbf{U} \geq 0, \mathbf{V} \geq 0, \mathbf{H} \geq 0, \mathbf{A} \geq 0 \end{aligned} \tag{12}$$

Where parameter $\alpha \in (0, 1)$ and $\beta = 1 - \alpha$, λ and γ are used to control the balance between the regularization terms and the error terms. $\mathbf{A} \in \mathbb{R}^{c \times r}$ is a non-negative matrix and is initialized randomly in our algorithm.

Obviously, GDJCF is a kind of novel supervised concept factorization. We explicitly incorporate the multi-features, the graph Laplacian and label information into the objective function of CF. In the formula (12), the first two items are the reconstruction errors using two kinds of features, the third term guarantees that two adjacent points in primitive space have similar representation in manifold space, the fourth term makes the learned representation have the ability to discriminate.

3.2   The Update Rules of GDJCF

Although the objective function in (12) is convex for $\mathbf{U}$ only, $\mathbf{V}$ only, $\mathbf{H}$ only or $\mathbf{A}$ only, it is not convex for four variables together. Thus, we extend the multiplicative update rules based on Euclidean distance of the original CF [26], aiming to find a local optimum, we have the following updating rules for $\mathbf{U}, \mathbf{V}, \mathbf{H}, \mathbf{A}$.

$$\mathbf{U}_{jk} \longleftarrow \mathbf{U}_{jk} \frac{(\mathbf{X}^T \mathbf{XH}^T)_{jk}}{(\mathbf{X}^T \mathbf{XUHH}^T)_{jk}} \tag{13}$$

$$\mathbf{V}_{jk} \longleftarrow \mathbf{V}_{jk} \frac{(\mathbf{Y}^T \mathbf{YH}^T)_{jk}}{(\mathbf{Y}^T \mathbf{YVHH}^T)_{jk}} \tag{14}$$

$$\mathbf{H}_{kj} \longleftarrow \mathbf{H}_{kj} \frac{(\alpha \mathbf{U}^T \mathbf{X}^T \mathbf{X} + \beta \mathbf{V}^T \mathbf{Y}^T \mathbf{Y} + \lambda \mathbf{HC} + \gamma \mathbf{A}^T \mathbf{S})_{kj}}{(\alpha \mathbf{U}^T \mathbf{X}^T \mathbf{X} \mathbf{U} \mathbf{H} + \beta \mathbf{V}^T \mathbf{Y}^T \mathbf{Y} \mathbf{V} \mathbf{H} + \lambda \mathbf{HB} + \gamma \mathbf{A}^T \mathbf{A} \mathbf{H})_{kj}} \tag{15}$$

$$\mathbf{A}_{qk} \longleftarrow \mathbf{A}_{qk} \frac{(\mathbf{SH}^T)_{qk}}{(\mathbf{AHH}^T)_{qk}} \tag{16}$$

We can iteratively update $\mathbf{U}$, $\mathbf{V}$, $\mathbf{H}$, and $\mathbf{A}$ until the objective function value does not change or the number of iteration exceed the maximum value. The procedure is summarized in Table 1.

Table 1. The algorithm of graph regularized discriminative joint concept factorization (GDJCF)

Input: Data matrix $\mathbf{X} \in \mathbb{R}^{m \times n}$, $\mathbf{Y} \in \mathbb{R}^{d \times n}$, graph Laplacian matrix $\mathbf{L} \in \mathbb{R}^{n \times n}$, indicator matrix $\mathbf{S} \in \mathbb{R}^{c \times n}$, parameters α, β, λ, γ
Initialization: Randomly initialize four non-negative matrices $\mathbf{U} \in \mathbb{R}^{n \times r}$, $\mathbf{V} \in \mathbb{R}^{n \times r}$, $\mathbf{H} \in \mathbb{R}^{r \times n}$ and $\mathbf{A} \in \mathbb{R}^{c \times r}$

Repeat
    1. Update $\mathbf{U}$ by rule (13)
    2. Update $\mathbf{V}$ by rule (14)
    3. Update $\mathbf{H}$ by rule (15)
    4. Update $\mathbf{A}$ by rule (16)
Until Convergence

Output: Coefficient matrix $\mathbf{H}$

---

4   Experimental Results

In this section, we first introduce relevant image data sets, and then illustrate our experiment settings, finally compare our scheme with other classical methods on five image data sets, i.e., ORL, Coil20, UMIST, Georgia Tech face, and UIUC texture data sets.

4.1   Data Sets

Five image data sets are used in the experiment. Three of them are face data sets, the another one is object image data and the last one is the texture image data. The important statistics of these data sets are summarized in Table 2. The examples of images from five data sets are shown in Fig. 1.

The ORL face data set[1] consists 400 images of 40 different subjects in PGM format. Each subject has 10 images. Subjects were asked to face the camera and no restrictions were imposed on expression; only limited side movement and limited tilt were tolerated. For most subjects, the images were shot at different times and with different lighting conditions, but all the images were taken against a dark homogeneous background. Some subjects were captured with and without glass.

---
[1] http://www.cl.cam.ac.uk/research/dtg/attarchive/facedatabase.html.

Table 2. Statistics of the five image data sets

Data set	Number of samples/classes	Dimensionality of gray value/lbp
ORL	400/40	256/256
Coil20	1440/20	256/256
UMIST	575/20	256/256
Georgia Tech face	750/50	256/256
UIUC texture	1000/25	256/256

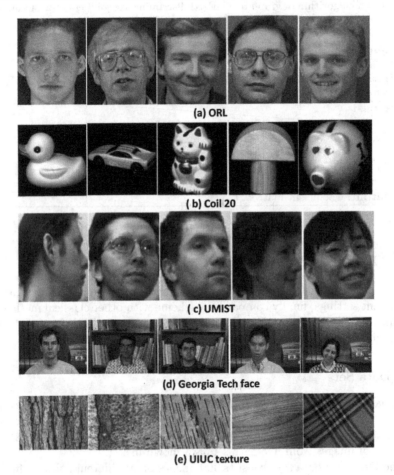

(a) ORL

(b) Coil 20

(c) UMIST

(d) Georgia Tech face

(e) UIUC texture

Fig. 1. Examples of images from five data sets

The Coil20 object data set[2] consists 1440 images of 20 different classes in PNG format. Each class has 72 images and each image contains both the object and the black background.

---

[2] http://www.cs.columbia.edu/CAVE/software/softlib/coil-20.php.

The UMIST face data set[3] consists 575 images of 20 individuals (mixed race/gender/appearance). Each individual is shown in a range of poses from profile to frontal views. The images are all in PGM format, approximately 220×220 pixels with 256-bit grey scale.

The Georgia Tech face data set[4] consists 750 images of 50 different classes in JPEG format. For each class, there are 15 color images. Most of the images were taken in two different sessions to take into account the variations in illumination conditions, facial expression, and appearance. In addition to this, the faces were captured at different scales and orientations.

The UIUC texture data set[5] consists 1000 images of 25 classes. All images are in gray scale JPG format, 640×480 pixels.

4.2   Experiment Settings

In the experiment, we extract gray value and LBP feature simultaneously. Specifically, we employ $\mathbf{X} \in \mathbb{R}^{m \times n}$ and $\mathbf{Y} \in \mathbb{R}^{d \times n}$ to indicate gray value and LBP feature extracted from images respectively. Combining the above two kinds features, we can get $\mathbf{Z} = [\mathbf{X}; \mathbf{Y}] \in \mathbb{R}^{p \times n}$. After obtaining the feature combination $\mathbf{Z}$, we put it into formula (1), (2), (4), (6) and (8) instead of $\mathbf{X}$.

In the our GDJCF algorithm, there are four parameters, i.e., α, β, λ and γ. The best empirical choice is $\alpha = 0.8$, $\beta = 0.2$, $\lambda = 4$ and $\gamma = 0.1$. For fair comparison, parameters appeared in all algorithms are set as the same. Besides, we set iteration number equal to 2000 when using multiplicative update rules. In the GNMF, LCCF and GDJCF, for each data point $\mathbf{z}_j$, we find its 5 nearest neighbors and put edges between $\mathbf{z}_j$ and its neighbors, 0–1 weighting function is used to define the weight matrix on the graph. In order to randomize the experiments, we conduct the evaluations with different cluster numbers. For each given cluster number K, 20 test runs are conducted on different randomly chosen clusters (except the case when the entire data set is used). The mean and standard error of the performance are reported in the Tables. All experiments are conducted in MATLAB 2011, which is executed on a PC with an Intel Core i7-4790 CPU (3.60 GHz) and 8 GB RAM.

4.3   Comparative Analysis

The clustering result is evaluated by comparing the obtained label of each sample with the label provided by the data set. Two metrics, the accuracy (AC) and the normalized mutual information (NMI) are used to measure the clustering performance. Please see [4] for the detailed definitions of these two metrics.

Tables 3, 4, 5, 6, 7, 8, 9, 10, 11 and 12 show the clustering results on the five image data sets. From the Tables, we can see that our GDJCF scheme always outperform the other five classical algorithms except on the ORL and Coil20 data sets, and the best results are marked in bold fonts.

---

[3] https://www.sheffield.ac.uk/eee/research/iel/research/face.

[4] http://www.face-rec.org/databases/.

[5] http://www-cvr.ai.uiuc.edu/ponce_grp/data/.

Table 3. Clustering performance on ORL (ACC)

K	Accuracy (%)					
	K-Means	NMF	GNMF	CF	LCCF	GDJCF
5	92.2 ± 11.9	88.6 ± 11.6	92.4 ± 11.8	86.1 ± 12.8	90.1 ± 12.0	**95.3 ± 8.8**
10	79.4 ± 8.8	80.9 ± 7.6	84.7 ± 5.7	76.8 ± 7.9	82.2 ± 7.0	**85.2 ± 5.8**
15	77.2 ± 4.6	78.3 ± 7.5	79.1 ± 6.1	70.6 ± 5.6	78.3 ± 6.7	**81.4 ± 4.3**
20	75.4 ± 4.8	78.1 ± 4.8	80.0 ± 4.4	71.4 ± 5.1	77.0 ± 4.4	**81.0 ± 6.1**
25	72.7 ± 4.5	75.2 ± 3.8	**76.4 ± 3.3**	67.4 ± 4.0	75.1 ± 4.3	75.2 ± 2.6
30	70.9 ± 3.9	72.6 ± 3.4	73.6 ± 3.8	65.5 ± 3.9	73.2 ± 4.0	**75.7 ± 5.7**
35	70.2 ± 3.6	71.3 ± 4.1	72.9 ± 3.3	63.0 ± 3.4	72.5 ± 2.9	**75.4 ± 4.0**
40	68.8	75.0	70.3	66.8	71.8	**74.3**

Table 4. Clustering performance on ORL (NMI)

K	Normalized mutual information (%)					
	K-Means	NMF	GNMF	CF	LCCF	GDJCF
5	92.1 ± 10.6	86.4 ± 12.9	92.3 ± 10.4	87.7 ± 10.5	90.0 ± 11.5	**95.9 ± 6.8**
10	85.9 ± 6.2	85.9 ± 5.8	89.5 ± 4.2	83.1 ± 5.3	88.1 ± 4.4	**93.0 ± 2.5**
15	86.0 ± 3.4	86.5 ± 4.3	87.8 ± 3.3	81.9 ± 3.5	86.6 ± 3.8	**91.6 ± 2.3**
20	86.4 ± 2.8	87.2 ± 2.6	89.1 ± 2.7	82.3 ± 3.4	86.6 ± 2.3	**91.9 ± 3.5**
25	85.2 ± 2.5	86.5 ± 2.4	87.7 ± 2.2	81.2 ± 2.5	86.4 ± 2.0	**88.9 ± 2.2**
30	84.9 ± 1.6	85.9 ± 1.7	87.2 ± 1.6	80.3 ± 2.3	86.4 ± 1.5	**89.3 ± 3.9**
35	84.8 ± 1.6	85.0 ± 2.3	86.7 ± 1.5	79.3 ± 1.7	85.6 ± 1.4	**89.4 ± 2.8**
40	83.7	86.8	84.5	79.6	86.7	**87.8**

Table 5. Clustering performance on Coil20 (ACC)

K	Accuracy (%)					
	K-Means	NMF	GNMF	CF	LCCF	GDJCF
2	90.1 ± 15.4	86.4 ± 17.9	87.1 ± 18.7	84.6 ± 19.2	92.8 ± 15.8	**96.7 ± 11.1**
4	76.7 ± 16.9	82.2 ± 15.8	93.9 ± 10.1	77.1 ± 13.2	90.2 ± 13.6	**96.8 ± 5.9**
6	71.3 ± 10.8	74.0 ± 14.3	87.4 ± 11.8	70.9 ± 14.3	86.6 ± 9.5	**90.7 ± 10.8**
8	71.7 ± 8.0	72.2 ± 7.4	91.1 ± 10.3	69.6 ± 10.0	85.8 ± 11.0	**91.3 ± 9.3**
10	71.9 ± 4.8	74.8 ± 9.7	82.8 ± 8.5	69.3 ± 8.9	81.4 ± 4.8	**84.3 ± 7.0**
12	67.3 ± 6.3	69.4 ± 8.6	80.5 ± 7.0	61.5 ± 6.9	79.3 ± 5.6	**83.0 ± 6.0**
14	66.1 ± 4.5	67.8 ± 5.0	79.9 ± 3.8	60.3 ± 4.0	79.4 ± 4.5	**80.6 ± 4.8**
16	64.7 ± 3.7	68.0 ± 5.7	79.5 ± 5.0	56.9 ± 4.5	76.1 ± 3.3	**81.1 ± 4.5**
18	64.0 ± 3.0	67.3 ± 4.0	**79.2 ± 5.3**	53.7 ± 4.4	74.8 ± 4.8	77.8 ± 5.3
20	63.8	70.1	72.1	51.7	79.0	**84.1**

Table 6. Clustering performance on Coil20 (NMI)

K	(%)					
	K-Means	NMF	GNMF	CF	LCCF	GDJCF
2	73.4 ± 37.7	65.2 ± 42.1	68.8 ± 42.6	62.4 ± 44.0	82.7 ± 35.1	$\mathbf{91.6 \pm 23.9}$
4	69.6 ± 18.9	75.7 ± 19.0	91.8 ± 11.1	69.1 ± 15.8	88.1 ± 15.0	$\mathbf{94.3 \pm 8.6}$
6	70.8 ± 12.8	73.7 ± 14.3	90.0 ± 8.7	70.6 ± 16.3	88.0 ± 7.3	$\mathbf{91.0 \pm 8.7}$
8	75.1 ± 6.9	75.0 ± 7.6	$\mathbf{92.9 \pm 6.5}$	70.3 ± 9.6	88.3 ± 7.9	92.1 ± 7.8
10	76.5 ± 5.0	78.2 ± 8.4	88.6 ± 6.2	72.4 ± 7.6	87.1 ± 3.9	$\mathbf{88.7 \pm 5.5}$
12	74.2 ± 5.3	75.3 ± 6.1	87.4 ± 4.9	67.8 ± 6.6	86.1 ± 4.6	$\mathbf{88.7 \pm 4.4}$
14	73.9 ± 3.4	75.8 ± 4.4	$\mathbf{87.9 \pm 2.4}$	66.5 ± 3.5	86.3 ± 3.3	87.9 ± 3.4
16	73.9 ± 2.9	75.9 ± 4.1	$\mathbf{88.0 \pm 2.6}$	64.2 ± 4.2	85.0 ± 2.2	87.9 ± 3.2
18	74.4 ± 2.4	75.8 ± 2.7	$\mathbf{87.7 \pm 2.5}$	63.5 ± 3.4	84.5 ± 2.6	86.5 ± 3.3
20	76.0	78.1	87.0	62.5	86.3	$\mathbf{89.4}$

Table 7. Clustering performance on UMIST (ACC)

K	Accuracy (%)					
	K-Means	NMF	GNMF	CF	LCCF	GDJCF
4	64.0 ± 9.3	67.5 ± 10.9	78.7 ± 15.2	62.5 ± 10.2	71.3 ± 13.0	$\mathbf{95.9 \pm 8.4}$
6	55.3 ± 6.6	61.8 ± 7.9	69.6 ± 13.2	58.2 ± 7.0	64.9 ± 11.1	$\mathbf{95.4 \pm 7.8}$
8	54.5 ± 7.7	57.0 ± 7.8	62.4 ± 6.3	54.9 ± 5.2	57.9 ± 8.4	$\mathbf{89.6 \pm 9.8}$
10	50.3 ± 4.2	52.1 ± 5.9	56.7 ± 7.1	49.4 ± 4.6	58.3 ± 5.9	$\mathbf{83.2 \pm 5.7}$
12	48.6 ± 4.8	52.5 ± 5.0	57.4 ± 5.6	48.6 ± 4.7	53.7 ± 3.9	$\mathbf{79.0 \pm 9.7}$
14	46.0 ± 4.5	48.8 ± 5.1	52.7 ± 5.6	48.0 ± 4.7	50.5 ± 5.0	$\mathbf{79.3 \pm 7.2}$
16	46.3 ± 2.1	47.8 ± 4.1	52.4 ± 3.1	45.8 ± 3.6	50.4 ± 4.3	$\mathbf{77.1 \pm 7.7}$
18	44.0 ± 2.3	47.7 ± 3.2	50.3 ± 3.2	45.4 ± 3.2	49.2 ± 3.1	$\mathbf{75.7 \pm 9.4}$
20	45.2	43.7	49.4	46.6	44.9	$\mathbf{69.6}$

Table 8. Clustering performance on UMIST (NMI)

K	Normalized mutual information (%)					
	K-Means	NMF	GNMF	CF	LCCF	GDJCF
4	56.8 ± 13.7	56.7 ± 13.9	74.1 ± 18.4	52.7 ± 13.2	65.7 ± 17.0	$\mathbf{95.3 \pm 8.6}$
6	56.9 ± 6.7	64.4 ± 7.8	72.9 ± 11.0	58.5 ± 6.8	67.8 ± 10.1	$\mathbf{97.0 \pm 4.8}$
8	60.9 ± 5.8	64.4 ± 5.7	70.7 ± 7.1	59.4 ± 4.9	65.2 ± 7.4	$\mathbf{93.2 \pm 5.6}$
10	60.4 ± 4.5	62.0 ± 6.5	68.5 ± 6.8	60.2 ± 5.0	68.4 ± 5.7	$\mathbf{90.2 \pm 3.9}$
12	61.8 ± 3.5	64.3 ± 5.5	70.0 ± 3.9	60.8 ± 4.4	66.9 ± 3.3	$\mathbf{88.2 \pm 6.0}$
14	61.5 ± 3.4	63.4 ± 4.2	68.6 ± 4.0	62.6 ± 3.5	65.8 ± 3.1	$\mathbf{88.5 \pm 4.9}$
16	62.7 ± 1.7	63.5 ± 2.8	68.6 ± 2.8	61.9 ± 3.3	67.0 ± 1.9	$\mathbf{87.4 \pm 4.6}$
18	62.8 ± 2.4	64.4 ± 2.7	68.1 ± 2.2	63.4 ± 2.8	67.0 ± 1.9	$\mathbf{86.2 \pm 5.6}$
20	63.2	63.4	69.1	62.7	65.5	$\mathbf{84.3}$

Table 9. Clustering performance on Georgia Tech face (ACC)

K	Accuracy (%)					
	K-Means	NMF	GNMF	CF	LCCF	GDJCF
5	70.7 ± 10.2	65.5 ± 10.2	72.0 ± 11.8	65.2 ± 11.4	70.5 ± 11.9	$\mathbf{86.0 \pm 15.3}$
10	65.1 ± 7.6	67.1 ± 7.8	67.0 ± 7.5	60.6 ± 8.5	64.2 ± 7.9	$\mathbf{86.2 \pm 11.4}$
15	61.1 ± 5.8	64.7 ± 5.0	65.2 ± 7.3	55.5 ± 5.7	60.9 ± 4.2	$\mathbf{76.5 \pm 6.0}$
20	59.6 ± 5.1	61.7 ± 3.9	63.7 ± 5.1	50.9 ± 4.7	58.9 ± 5.1	$\mathbf{80.2 \pm 6.0}$
25	56.2 ± 4.0	60.4 ± 3.8	61.2 ± 4.0	49.7 ± 3.7	57.2 ± 3.7	$\mathbf{74.1 \pm 5.2}$
30	55.7 ± 4.3	59.9 ± 4.9	61.2 ± 2.6	46.7 ± 3.8	54.5 ± 3.9	$\mathbf{73.8 \pm 4.4}$
35	53.3 ± 3.1	57.1 ± 2.6	60.1 ± 3.0	44.2 ± 3.4	53.0 ± 3.3	$\mathbf{72.8 \pm 5.5}$
40	54.0 ± 2.1	55.2 ± 3.1	57.8 ± 3.4	42.1 ± 2.1	53.1 ± 2.9	$\mathbf{72.5 \pm 4.7}$
45	52.2 ± 2.1	53.1 ± 2.0	57.2 ± 2.5	40.8 ± 2.1	51.1 ± 1.7	$\mathbf{71.2 \pm 3.8}$
50	54.5	50.8	56.0	38.4	51.3	**71.2**

Table 10. Clustering performance on Georgia Tech face (NMI)

K	Normalized mutual information (%)					
	K-Means	NMF	GNMF	CF	LCCF	GDJCF
5	67.5 ± 10.7	62.6 ± 11.3	69.1 ± 11.6	58.3 ± 11.2	66.9 ± 12.5	$\mathbf{87.8 \pm 13.5}$
10	68.1 ± 6.3	70.1 ± 7.5	71.8 ± 6.8	62.2 ± 7.6	68.6 ± 7.1	$\mathbf{91.2 \pm 8.6}$
15	68.7 ± 4.7	71.2 ± 4.3	72.8 ± 5.1	62.7 ± 4.9	69.4 ± 3.9	$\mathbf{86.4 \pm 5.3}$
20	69.9 ± 4.1	71.3 ± 3.8	73.9 ± 3.9	61.9 ± 3.5	69.9 ± 4.3	$\mathbf{89.2 \pm 5.1}$
25	69.1 ± 2.9	71.7 ± 3.1	73.8 ± 2.5	62.7 ± 2.5	70.1 ± 3.0	$\mathbf{86.5 \pm 4.6}$
30	69.9 ± 2.9	72.3 ± 3.3	74.6 ± 1.6	61.9 ± 2.7	69.9 ± 2.6	$\mathbf{86.6 \pm 3.4}$
35	69.2 ± 2.3	71.6 ± 1.6	74.3 ± 2.0	61.2 ± 2.1	69.3 ± 2.7	$\mathbf{86.4 \pm 4.4}$
40	69.8 ± 1.7	70.6 ± 2.1	73.7 ± 2.1	60.7 ± 1.7	70.0 ± 2.2	$\mathbf{85.6 \pm 4.3}$
45	69.7 ± 1.3	70.2 ± 1.8	73.4 ± 1.5	60.5 ± 1.3	69.5 ± 1.2	$\mathbf{85.5 \pm 3.2}$
50	71.5	69.3	73.9	60.3	70.1	**86.3**

Table 11. Clustering performance on UIUC texture (ACC)

K	Accuracy (%)					
	K-Means	NMF	GNMF	CF	LCCF	GDJCF
2	78.3 ± 14.0	77.4 ± 12.3	77.9 ± 13.8	77.7 ± 12.4	78.4 ± 12.6	$\mathbf{91.9 \pm 12.2}$
4	51.7 ± 6.9	50.3 ± 6.9	49.6 ± 5.8	49.2 ± 6.5	51.6 ± 7.0	$\mathbf{84.9 \pm 17.1}$
6	42.2 ± 4.7	43.8 ± 5.7	46.8 ± 6.5	41.6 ± 5.0	42.9 ± 5.4	$\mathbf{73.6 \pm 16.4}$
8	38.5 ± 4.9	43.4 ± 5.4	42.1 ± 5.8	38.7 ± 5.0	38.5 ± 4.6	$\mathbf{69.8 \pm 17.8}$
10	34.5 ± 2.5	39.9 ± 5.2	37.8 ± 3.4	34.0 ± 2.1	36.7 ± 4.1	$\mathbf{65.3 \pm 14.2}$
12	32.4 ± 2.8	37.0 ± 4.3	35.5 ± 4.1	33.7 ± 3.3	35.8 ± 3.8	$\mathbf{67.6 \pm 16.8}$
14	30.2 ± 2.8	38.2 ± 3.4	34.3 ± 3.1	33.7 ± 2.9	35.2 ± 3.1	$\mathbf{70.4 \pm 9.7}$
16	28.3 ± 1.9	36.7 ± 3.4	32.5 ± 2.9	31.6 ± 2.4	33.2 ± 1.8	$\mathbf{67.4 \pm 10.5}$
18	27.4 ± 1.5	35.3 ± 2.0	32.1 ± 2.3	31.2 ± 2.6	33.1 ± 2.1	$\mathbf{68.3 \pm 7.9}$
20	26.1 ± 1.7	33.3 ± 2.5	30.0 ± 2.5	30.6 ± 1.4	31.4 ± 1.8	$\mathbf{64.5 \pm 6.8}$
22	26.0 ± 1.6	32.2 ± 1.4	29.7 ± 1.9	30.3 ± 1.9	30.9 ± 1.8	$\mathbf{61.7 \pm 9.5}$
25	24.4	33.3	29.1	29.0	30.1	**55.4**

Table 12. Clustering performance on UIUC texture (NMI)

K	Normalized mutual information (%)					
	K-Means	NMF	GNMF	CF	LCCF	GDJCF
2	36.4 ± 30.3	32.7 ± 26.0	35.3 ± 30.0	33.3 ± 25.3	35.8 ± 27.3	$\mathbf{75.7 \pm 35.2}$
4	30.1 ± 9.1	26.9 ± 8.8	27.4 ± 8.4	25.3 ± 8.7	28.6 ± 9.0	$\mathbf{84.2 \pm 18.8}$
6	31.6 ± 5.8	32.7 ± 6.6	34.9 ± 7.0	30.7 ± 6.3	28.5 ± 6.2	$\mathbf{74.6 \pm 19.2}$
8	33.7 ± 4.6	39.3 ± 5.4	36.5 ± 6.4	34.9 ± 5.0	31.6 ± 6.4	$\mathbf{73.6 \pm 20.3}$
10	32.8 ± 3.2	39.2 ± 5.2	38.0 ± 3.2	34.9 ± 3.7	36.3 ± 5.1	$\mathbf{71.8 \pm 17.4}$
12	33.7 ± 2.8	39.3 ± 3.7	37.7 ± 4.0	37.4 ± 3.7	37.8 ± 4.3	$\mathbf{74.2 \pm 18.9}$
14	34.0 ± 2.6	42.6 ± 3.3	38.3 ± 3.5	39.1 ± 2.3	40.2 ± 3.4	$\mathbf{80.4 \pm 7.4}$
16	34.4 ± 1.9	42.4 ± 3.6	38.6 ± 3.0	39.1 ± 2.4	39.9 ± 2.1	$\mathbf{78.9 \pm 8.2}$
18	34.4 ± 1.5	42.4 ± 2.5	39.5 ± 2.6	39.7 ± 2.6	41.4 ± 2.5	$\mathbf{80.5 \pm 6.0}$
20	35.1 ± 1.7	42.2 ± 2.1	39.5 ± 2.4	41.0 ± 1.5	41.3 ± 1.5	$\mathbf{77.1 \pm 6.1}$
22	36.0 ± 1.7	42.7 ± 1.4	40.2 ± 2.0	41.7 ± 1.7	42.5 ± 1.9	$\mathbf{75.9 \pm 9.2}$
25	35.0	44.3	41.2	41.2	43.2	$\mathbf{69.8}$

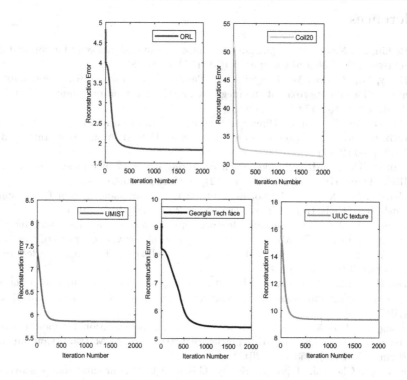

Fig. 2. The convergence validations of GDJCF algorithm on the five image data sets

4.4   Convergence Analysis

We test the convergence of GDJCF algorithm on the five image data sets. The reconstruction error versus the iteration number is as following Fig. 2. Our method converges after about 500 iterations.

5   Conclusions

In this paper, inspired by the joint non-negative matrix factorization, manifold learning theory and label information, the GDJCF algorithm is proposed. Experimental results reflect the effectiveness of GDJCF. In our future research, we will combine the joint learning with some variants of CF and test on more data sets.

Acknowledgment. This work is supported in part by the Natural Science Foundation of Jiangsu Province (Grant No. BK20150856), the NUPTSF (Grant No. NY215144), the Postdoctoral Research Plan of Jiangsu Province (Grant No. 1501054B), the Post-doctoral Science Foundation of China (Grant No. 2016M591840).

References

1. Belkin, M., Niyogi, P.: Laplacian eigenmaps for dimensionality reduction and data representation. Neural Comput. **15**(6), 1373–1396 (2003)
2. Berry, M., Browne, M., Langville, A., Pauca, V., Plemmons, R.: Algorithms and applications for approximate nonnegative matrix factorization. Comput. Stat. Data Anal. **52**(1), 155–173 (2007)
3. Cai, D., He, X., Han, J., Huang, T.S.: Graph regularized non-negative matrix factorization for data representation. IEEE Trans. Pattern Anal. Mach. Intell. **33**(8), 1548–1560 (2011)
4. Cai, D., He, X., Han, J.: Document clustering using locality preserving indexing. IEEE Trans. Knowl. Data Eng. **17**(12), 1624–1637 (2005)
5. Cai, D., He, X., Han, J.: Locally consistent concept factorization for document clustering. IEEE Trans. Knowl. Data Eng. **23**(6), 902–913 (2011)
6. Coates, A., Ng, A.Y.: Learning feature representations with K-means. In: Montavon, G., Orr, G.B., Müller, K.-R. (eds.) Neural Networks: Tricks of the Trade. LNCS, vol. 7700, pp. 561–580. Springer, Heidelberg (2012). https://doi.org/10.1007/978-3-642-35289-8_30
7. Du, R., Drake, B., Park, H.: Hybrid clustering based on content and connection structure using joint nonnegative matrix factorization. J. Glob. Optim. **33**(6), 2017 (2017)
8. Jiang, X., Hu, X., Xu, W.: Microbiome data representation by joint nonnegative matrix factorization with laplacian regularization. ACM Trans. Comput. Biol. Bioinform. **14**(2), 353–359 (2017)
9. Kim, H., Choo, J., Kim, J., Reddy, C.K., Park, H.: Simultaneous discovery of common and discriminative topics via joint nonnegative matrix factorization. In: The ACM SIGKDD International Conference, pp. 567–576 (2015)
10. Lee, D., Seung, H.: Algorithms for non-negative matrix factorization. In: Advances in Neural Information Processing Systems, vol. 13 (2001)

11. Lee, D., Seung, H., et al.: Learning the parts of objects by non-negative matrix factorization. Nature **401**(6755), 788–791 (1999)
12. Li, H., Zhang, J., Hu, J., Zhang, C., Liu, J.: Graph-based discriminative concept factorization for data representation. Soft. Comput. **118**(C), 1–13 (2016)
13. Li, H., Zhang, J., Liu, J.: Class-driven concept factorization for image representation. Neurocomputing **190**, 197–208 (2016)
14. Li, X., Shen, X., Shu, Z., Ye, Q., Zhao, C.: Graph regularized multilayer concept factorization for data representation. Neurocomputing **238**(C), 139–151 (2017)
15. Liu, G., Lin, Z., Yan, S., Sun, J., Yu, Y., Ma, Y.: Robust recovery of subspace structures by low-rank representation. IEEE Trans. Pattern Anal. Mach. Intell. **35**(1), 171–184 (2013)
16. Liu, H., Wu, Z., Cai, D., Huang, T.S.: Constrained nonnegative matrix factorization for image representation. IEEE Trans. Pattern Anal. Mach. Intell. **34**(7), 1299–1311 (2012)
17. Liu, J., Wang, C., Gao, J., Han, J.: Multi-view clustering via joint nonnegative matrix factorization. In: Proceedings of the 2013 SIAM International Conference on Data Mining, pp. 252–260 (2013)
18. Lu, M., Zhao, X.J., Zhang, L., Li, F.Z.: Semi-supervised concept factorization for document clustering. Inf. Sci. **331**(C), 86–98 (2016)
19. Luo, X., Zhou, M., Xia, Y., Zhu, Q.: An efficient non-negative matrix-factorization-based approach to collaborative filtering for recommender systems. IEEE Trans. Industr. Inf. **10**(2), 1273–1284 (2014)
20. Pei, X., Chen, C., Gong, W.: Concept factorization with adaptive neighbors for document clustering. IEEE Trans. Neural Netw. Learn. Syst. 1–10 (2016)
21. Peng, Y., Lu, B.L.: Discriminative extreme learning machine with supervised sparsity preserving for image classification. Neurocomputing **261**, 242–252 (2017)
22. Song, P., et al.: Transfer semi-supervised non-negative matrix factorization for speech emotion recognition. IEICE Trans. Inf. Syst. **99**, 2647–2650 (2016)
23. Wang, Y.X., Zhang, Y.J.: Nonnegative matrix factorization: a comprehensive review. IEEE Trans. Knowl. Data Eng. **25**(6), 1336–1353 (2013)
24. Wright, J., Yang, A.Y., Ganesh, A., Sastry, S.S., Ma, Y.: Robust face recognition via sparse representation. IEEE Trans. Pattern Anal. & Mach. Intell. **31**(2), 210 (2009)
25. Wu, Z., Chng, E.S., Li, H.: Joint nonnegative matrix factorization for exemplar-based voice conversion. Multimed. Tools Appl. **74**(22), 9943–9958 (2014)
26. Xu, W., Gong, Y.: Document clustering by concept factorization. In: International ACM SIGIR Conference on Research and Development in Information Retrieval, pp. 202–209 (2004)
27. Zhang, T., Fang, B., Tang, Y.Y., He, G., Wen, J.: Topology preserving non-negative matrix factorization for face recognition. IEEE Trans. Image Process. **17**(4), 574–584 (2008)

Influence Maximization Node Mining with Trust Propagation Mechanism

Hui Zhang[1(✉)], DongZhi Wang[1], ChunMing Yang[1], XuJian Zhao[1], Bo Li[1], and Fei Yuan[2]

[1] School of Computer Science and Technology,
Southwest University of Science and Technology, 621010 Mianyang, China
zhanghui@swust.edu.cn
[2] Big Data Research Center,
University of Electronic Science and Technology of China,
611731 Chengdu, China

Abstract. The issue of maximizing influence is one of the hot issues in the study of complex networks. It's of great significance for understanding the dissemination mechanism of network information and controlling rumor. In recent years, based on the percolation theory, the problem of maximizing the node identification has attracted a lot of attention. However, this method does not consider the influence of the propagation of trust on the maximization of influence. This paper introduces the trust transfer function to depict the phenomenon that trust value and distrust value decreases and increase, respectively, and uses the percolation theory to solve the node joint propagation strength index to excavate the influencer. Experiments show that the proposed algorithm outperforms other heuristic benchmark algorithms.

Keywords: Complex network · Percolation theory · Trust propagation

1 Introduction

Influence analysis is an important part of social network analysis. As the basic work of influence analysis, the influence maximization node mining aims at distinguishing k nodes from the network. Under some influence propagation mechanism, these k nodes produce greatest influence spread range. The actual social network involves a large number of nodes and sparse network. Therefore, the maximization of influence mining under the social network is confronted with big challenge.

To improve algorithm execution, scholars identify the influence maximization node by heuristic algorithms. Pastorsatorras et al. [1–3] used HD (High-Degree) to measure

The next generation Internet technology innovation project (NGII20170901) of the SEIR network (NGII20170901); the open fund of the Sichuan Academy of military and civil integration (18sx b017, 18sxb028); the humanities and social sciences fund of the Ministry of Education (17YJCZ H260); the fund project of the Sichuan information management and service research center (SCT Q2016YB13).

Z.-H. Zhou et al. (Eds.): ICAI 2018, CCIS 888, pp. 84–97, 2018.
https://doi.org/10.1007/978-981-13-2122-1_7

the node importance. In other words, the node influence intensity is defined according to node degree value. Colizza [4, 5] proved that the highest node of degree value is strategically removed as influence node to spread the influence throughout the entire network in the "rich club" network. On this basis, Cohen [3] proposed HDA (High Degree Adaptive), which continuously removes the node with the highest degree and its edges as influence nodes from the network. Besides, it updates the degree order of all nodes. Bavelas [6] used CC (Closeness Centrality) to reflect the node centrality in the network, namely the reciprocal of sum of the shortest distances from Node i to all other nodes multiplied by the number of other nodes. The K-core of a network refers to the remaining subgraph after repeatedly removing the nodes with degree values smaller than k and the connected edges. Kitsak [7] proved that the K-core algorithm is suitable for social networks with independent maximum influence nodes. However, the K-core algorithm is extremely inefficient when the maximum influence nodes are jointly transmitted [8]. Therefore, the K-core algorithm is not optimal under the joint propagation social network. PageRank algorithm [9] measures the importance of web pages, the core of the algorithm is the feature vector centrality of web page connection. However, if a non-influential node is connected to the most influential node, then the node can obtain a higher "score" of influence. Thus, it has a high probability of being excavated as the next pseudo-influence maximization node.

Due to the limitations of heuristic algorithm, the set of maximum influence nodes is excavated from the information propagation modeling. Kempe et al. [8] modeled influence propagation as a time-discrete propagation process. The problem is formalized to a discrete optimization problem, such as LTM (linear threshold model) [10] and ICM (independent cascade model) [11], which are improved by massive jobs. For large-scale online social networks, Chen et al. [12] proved that it is NP-hard to calculate the expected value of each user's influence in a social network. In addition, the model is sensitive to the input parameters because of artificial definition of information diffusion threshold.

Morone and Makse [17] discussed the maximum influence problem according to the percolation theory, providing a new idea for the solution of this problem. Based on the traditional heuristic algorithm, a global influence optimization index is proposed by considering the joint transmission influence factors between influence nodes rather than the trust degree mechanism of influence transmission between nodes. On this basis, the trust transfer function is introduced to discuss the effects of the phenomenon of decreased trust and increased distrust degrees on the influence maximization in the dissemination process.

Artificial and real network datasets were used to evaluate the influence of algorithm. Compared with the common heuristic algorithm, we excavated the influence diffusivity of the set of influence maximization nodes to verify the rationality and effectiveness of the proposed algorithm.

2 Maximum Influence Node Mining Model Based on Percolation Theory

The work built the maximum influence node mining model from three angles.

1. Non-tracking matrix was used to build a social network diagram model that satisfies the information (influence) propagation conditions.
2. Trust transfer function was introduced into the information propagation model to build the actual influence propagation mechanism.
3. Fast maximum influence mining algorithm was obtained according to the percolation theory.

2.1 Social Network Diagram Model Based on Non-tracking Matrix

Recently, the non-tracking matrix has attracted much attention in the modeling of complex networks. It is denoted that an undirected social network has N nodes and M edges. The undirected graph is firstly transformed into directed graph to build a non-tracking matrix in the social network. The specific operation is described as follows. For two directly associated nodes i and j $(i, j \in N; ij \in E)$, the undirected edges are replaced by two directed edges with opposite directions. In other words, $i \to j$ and $j \to i$ are used to represent two directed edges. The elements of non-tracking matrix are defined as follows.

Definition 1 [13, 14]: It is denoted that the network have N nodes and M edges. Then Non-tracking Matrix B is a $2M \times 2M$ unsymmetrical matrix. The elements are as follows.

$$B_{k \to l, i \to j} = \begin{cases} 1 \; if \; l = i \; and \; j \neq k \\ 0 \; otherwise \end{cases}$$

Non-tracking matrix characterizes the feasible path of non-tracking walking, namely the influence propagation path. "Non-tracking" refers to a simple walk strategy where the return along the past path is not allowed.

2.2 Influence Propagation Mechanism Based on Trust

Traditionally, the propagation problem of influence in social networks is discussed by information cascade technology. Information cascading is a very common phenomenon in social networks. After the formation of information cascading, the bottom influence nodes easily affected by upper influence nodes make the same choices as previous individuals while ignoring their own opinions.

Figure 1 shows two independent first-order diffusion cascades. There is no diffusion cascade between nodes A and C. According to information diffusion theory, the influence gradually decreases with the social chain. The nodes on the bottom of diffusion chain are affected by the directly connected superior friend nodes. Meanwhile, these nodes also have indirect influence transmission with the high-level nodes. The current influence diffusion model cannot truly portray the transmission of influence in

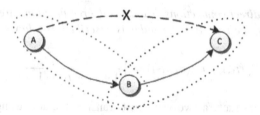

Fig. 1. Three-node information cascading

the social network, causing large error. Therefore, trust transfer function [16] is introduced to correctly simulate the influence propagation in social network.

Definition 2: *Node trust function. The trust degree function of each node is expressed as two-tuples* $\lambda = (t, d)$, *where* $t, s \in [0, 1]$. *For the trust transfer function, the first part is the trust degree of node; the second part the distrust degree of node. The trust function set in the social network is represented by* $\Lambda = \{\lambda = (t, d) | t, d \in [0, 1]\} \equiv [0, 1]^2$.

The essence of influence transmission mechanism in social network is the trust interaction between nodes. Nodes with higher trust degrees are more likely to achieve the maximization of influence. In influence transmission process, we introduced the concepts of triangular norms and conorms to portray the trust transfer mechanism between nodes [15]. If and only if function $T : [0, 1]^2 \rightarrow [0, 1]$ satisfies commutativity, associativity and monotonicity; boundary condition $T(x, 1) = x$ for $\forall x$, then Function T becomes a triangular norm. If and only if function $S : [0, 1]^2 \rightarrow [0, 1]$ satisfies commutativity, associativity and monotonicity; boundary condition $S(x, 0) = x$ for $\forall x$, then function S is a triangular conorm. In the work, the Einstein product $\otimes_\varepsilon$ is used as the triangular norm; the Einstein sum $\oplus_\varepsilon$ as the triangular conorm. The above two functions are regarded as trust transfer functions. For $\forall(a, b) \in [0, 1]^2$, there are the following equations.

$$E_\otimes = a \otimes_\varepsilon b = \frac{a \cdot b}{1 + (1 - a) \cdot (1 - b)} \tag{1}$$

$$E_\oplus = a \otimes_\varepsilon b = \frac{a + b}{1 + a \cdot b} \tag{2}$$

The triangular norm is the minimization operator; the triangular conorm the maximization operator. The above two operations have the following properties.

In the existing influence propagation model, the trust propagation mechanism does not involve the propagation of distrust degree in the influence propagation chain consisting of three or more nodes. More importantly, scholars do not consider the trust attenuation problem of influence propagation in the social network. There is a decrease in trust and an increase in distrust while the influence transmits in real social network. To achieve attenuation of trust and increase of distrust, we use Einstein product $\otimes_\varepsilon$ and Einstein summation $\oplus_\varepsilon$ as two-way trust transfer operators.

Definition 3: *Trust transfer index. It is denoted that* Λ *is the set of node trust functions. We construct two-way trust transfer operator* $P_D : \Lambda \times \Lambda \rightarrow \Lambda$, *which connects two* *voting nodes with direct connections. The trust degree functions are $\lambda_1 = (t_1, d_1)$ and $\lambda_2 = (t_2, d_2)$. Then, the trust transfer index is described as follows.*

$$P_D(\lambda_1, \lambda_2) = (E_\otimes(t_1, t_2), E_\oplus(d_1, d_2)) = \left(\frac{t_1 \cdot t_2}{1 + (1 - t_1) \cdot (1 - t_2)}, \frac{d_1 + d_2}{1 + d \cdot d_2} \right) \quad (3)$$

Definition 2 constructs a two-way trust transfer operator using $\otimes_\varepsilon$ and $\oplus_\varepsilon$. P_D satisfies $E_\otimes(t_1, t_2) \leq min\{t_1, t_2\}$ and $max\{d_1, d_2\} \leq E_\oplus(d_1, d_2)$. It realizes the trust transfer mechanism where the trust degree decreases, and the distrust degree increases.

The trust transfer operator P_D has the following properties.

- **Commutativity:**

$$\begin{aligned} P_D(\lambda_2, \lambda_1) &= (E_\otimes(t_1, t_2), E_\oplus(d_1, d_2)) \\ &= (E_\otimes(t_1, t_2), E_\oplus(d_1, d_2)) \\ &= P_D(\lambda_2, \lambda_1) \end{aligned}$$

- **Associativity:**

$$\begin{aligned} P_D(P_D(\lambda_1, \lambda_2), \lambda_3) &= P_D[(E_\otimes(t_1, t_2), E_\oplus(d_1, d_2)), (t_3, d_3)] \\ &= (E_\otimes(E_\otimes(t_1, t_2), t_3), E_\otimes(E_\otimes(d_1, d_2), d_3)) \\ &= (E_\otimes(t_1, E_\otimes(t_2, t_3)), E_\otimes(d_1, E_\otimes(d_2, d_3))) \\ &= P_D[(t_1, d_1), (E_\otimes(t_2, t_3), E_\oplus(d_2, d_3))] \\ &= P_D(\lambda_1, P_D(\lambda_2, \lambda_3)) \end{aligned}$$

- **Monotonicity:** $E_\otimes$ and $E_\oplus$ are monotonous, so P_D also satisfies monotony. With the spread of influence in the social chain, Trust Degree $P_D^{(1)}$ does not increase, and distrust degree $P_D^{(2)}$ is non-decreasing.
- **Boundary conditions:** According to the boundary conditions of the trust transfer operator P_D, the trust and distrust degrees are discussed, respectively.
  a. **Full trust transfer:** If $\lambda_1 = (1, 0)$, then $P_D((1, 0), \lambda_2) = (t_2, d_2) = \lambda_2$. If $\lambda_2 = (1, 0)$, then $P_D(\lambda_1, (1, 0)) = (t_1, d_1) = \lambda_1$ according to commutative laws. Therefore, when A has full confidence in B, the influence of B to C is completely dominated by A in an influence transfer chain with three nodes $\{A, B, C\} \in N$.
  b. **Total distrust transmission:** If $\lambda_1 = (0, 1)$, then $P_D((0, 1), \lambda_2) = (0, 1)$. According to commutativity, if $\lambda_2 = (0, 1)$, then $P_D(\lambda_1, (0, 1)) = (0, 1)$. Therefore, the node is not affected by any directly connected friend in a minimum connected subgraph composed of full distrust nodes.

The above illustrates the related properties of P_D. Figure 2 shows the trust transfer chain in the process of influence propagation with three nodes. Figure 2 shows the fully connected graph of three nodes, where there are actual social links between nodes. The trust transfer can directly calculate the trust and distrust degrees between nodes according to the above trust transfer operators. If the influence propagates in a linear chain (See Fig. 2), then it constructs an influence transfer chain from B→A→C. Actually, there are actual interaction and indirect trust transmission between B and C.

Therefore, connected dotted lines between B and C indicate that C is affected by second-order diffusion cascades of B.

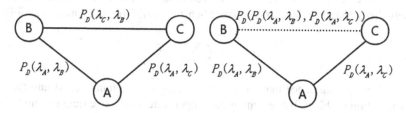

Fig. 2. The spread of trust in influence communication

Definition 4: *Influence diffusion coefficient. The influence diffusion coefficient between two connected nodes is the mapping* $P_D \rightarrow [0, 1]$ *on the trust transfer index. It is denoted that the trust degree functions of two nodes are* $v_1, v_2 \in V$ *and* $\lambda_1, \lambda_2 \in \Lambda$*; the trust transfer index between the nodes is* $P_D = \left(P_D^{(1)}, P_D^{(2)}\right)$*. Then, the influence diffusion coefficient is expressed as follows.*

$$IS(\lambda_1, \lambda_2) = \frac{1 - P_D^{(1)} + P_D^{(2)}}{2} \tag{4}$$

Through the trust transfer mechanism, we can calculate the influence diffusion coefficient between two nodes. This coefficient measures the probability that two active nodes cannot activate neighbor nodes in two directly connected nodes.

2.3 Maximum Influence Node Mining Under the Trust Degree Transfer

In the previous two sections, we established the social network representation model based on non-regressive matrix and the influence model in terms of social network propagation mechanism. After that, the maximum influence nodes are excavated by percolation theory. It is denoted that vector $n = (n_1, n_2, \ldots, n_N)$ represents the tag vector of influence node in the network. Wherein,

$$n_i = \begin{cases} 0 & n_i\ is\ removed\ node \\ 1 & n_i\ is\ present \end{cases}$$

The proportion of influence nodes in the network is:

$$q = 1 - \frac{1}{N} \sum_{i=1}^{N} n_i \equiv 1 - \langle n \rangle \tag{5}$$

Based on percolation theory, q_c "super-influence" nodes and their edges are constantly removed from the network nodes in the influence maximization node mining process. Therefore, the influence can no longer spread in the network. q_c is the least proportion of nodes to be removed from the network, namely the ratio of "super influence" nodes. It is denoted that $G(q)$ is the average probability of unaffected nodes after the influence diffusion in the network with q "super influence" nodes. It is denoted that $\mathbf{v} = (v_1, v_2, \ldots, v_N)$, where v_i is the probability that node i is finally unaffected (namely the probability that i is inactive when $t \to \infty$). $G(q)$ is expressed as follows.

$$G(q) = \frac{1}{N} \sum_{i=1}^{N} v_i \tag{6}$$

Therefore, the influence maximum node mining can be transformed into the optimal percolation problem. By searching the "super influence" node (removal node) with optimal q_c, we obtain minimum probability $G(q_c)$ of affected node in the final network. The mathematical form of problem is expressed as follows.

$$q_c = \min\left\{ q \in [0, 1] : \min G(q) \right\} \tag{7}$$

When $q \geq q_c$, there is a set of influence joint diffusion nodes in social network. The influence spreads from the nodes to the entire network. When $q \leq q_c$, there is a small isolated local area in the social network. Therefore, the influence cannot spread to the entire network.

To measure the actual influence of certain node in the network, the node is virtually removed to investigate the influence transfer changes. It is denoted that i and j have direct association; Tab $v_{i \to j}$ is the probability that i will not be affected when j is virtually removed. Taking i as the center, local influence diffusion tree is expressed as follows.

$$v_{i \to j} = n_i \left[1 - \prod_{k \in \partial i \backslash j} (1 - w_{k \to i} v_{k \to i}) \right] \tag{8}$$

$$w_{k \to i} = \frac{\sum_{t \in \partial k \backslash i} IS(\lambda_k, \lambda_t)}{|\partial k| - 1} \tag{9}$$

When i is an influence node, (i.e., $n_i = 0$), then $v_{i \to j} = 0$. When i is a non-influence node, (i.e., $n_i = 1$), then the probability whether the node is finally activated is related to the surrounding nodes. Wherein, $w_{k \to i}$ is the local average influence diffusion coefficient of node k when neighbor node i of node w is virtually removed; $\partial i \backslash j$ is the set of neighbor nodes of i after removing j virtually.

In the above local influence diffusion model, it can be clearly verified that $\{v_{i \to j} = 0\}$ is a globally stable solution for all $i \to j, i, j \in N$. According to the non-tracking matrix, the above model can be used to construct $2M \times 2M$ closable equations. The linear operator is introduced to solve the global optimum n.

$$\mathcal{M}_{k \to l, i \to j} = \left. \frac{\partial v_{i \to j}}{\partial v_{k \to l}} \right|_{\{v_{i \to j} = 0\}} \tag{10}$$

$\mathcal{M}$ is defined as the linear operator of local influence diffusion model in the $2M$ directed edges. This operator can be represented on a $2M \times 2M$ matrix based on the non-tracking matrix B. Therefore, weighted non-tracking matrix is defined after introducing influence diffusion coefficient in the diffusion model.

Definition 5: *Weighted non-tracking matrix under influence diffusion coefficient.*

$$\mathcal{M}_{k \to l, i \to j} = n_i w_{k \to i} \mathcal{B}_{k \to l, i \to j} \tag{11}$$

Based on percolation theory, the problem of influence maximum node mining is transformed into maximum eigenvalue problem of influence diffusion matrix M. According to the Perron-Frobenius theorem, the maximum eigenvalue of Matrix $w\mathcal{B}$ is strictly less than the function of $w\mathcal{B}$. When $\mathcal{M} \leq w\mathcal{B}$ (the element sizes of two matrices at the same position strictly satisfy the inequality requirement), then $\lambda(\mathcal{M}) < \lambda(w\mathcal{B})$. It is denoted that λ'_{max} is the maximum eigenvalue of $w\mathcal{B}$; λ_{max} the maximum eigenvalue of corresponding B; $\lambda(\boldsymbol{n}, q)$ the maximum eigenvalue of $\mathcal{M}$. Therefore, the problem is redefined as follows. We find the globally optimal influence node tag sequence $\boldsymbol{n}^*$ so that $\lambda(\boldsymbol{n}^*, q)$ reaches minimum, that is

$$\lambda(\boldsymbol{n}^*, q) \equiv \min_{\boldsymbol{n}:\langle n \rangle = 1-q} \lambda(\boldsymbol{n}, q) \tag{12}$$

The proportion q_c of optimal "super influence" nodes satisfies the following equation.

$$\lambda(\boldsymbol{n}^*, q_c) = \frac{\lambda_{max}}{\lambda'_{max}} \tag{.13}$$

It is a *NP*-hard problem to solve maximum eigenvalue of $\mathcal{M}$ by optimizing n. Using improved power method proposed in Reference [18], the maximum eigenvalue of matrix is explicitly expressed as the given joint influence propagation order ℓ.

$$\lambda_\ell(\boldsymbol{n}) = \left[\frac{|w_\ell(\boldsymbol{n})|}{|w_0|} \right]^{\frac{1}{\ell}} \tag{14}$$

$$|w_\ell(\boldsymbol{n})|^2 = \sum_{i=1}^N z_i \sum_{j \in Ball(i,\ell)} \left(\prod_{k \in P_\ell(i,j)} n_k \right) z_i \tag{15}$$

where $z_i = \sum_{t \in \partial i} w_{i \to t}$; $|w_0(\boldsymbol{n})|^2 = 2 \sum_{i,j} IS(\lambda_i, \lambda_j)$; $Ball(i, \ell)$ is the friend circle boundary node taking i as the center and ℓ as the radius; $P_\ell(i,j)$ the path node from i to friend circle boundary node j with the order of ℓ. Influence order ℓ measures the average reachable condition of joint propagation between two social nodes. Specifically, one of influence nodes needs to go through ℓ paths to co-propagate with another influence node.

Joint propagation between two "super influence" nodes i and j contributes to solving the above problem of minimizing the maximum eigenvalue, i.e., maximizing node influence. Then, there are ℓ paths between the two nodes ($n_k = 1, k \in P_\ell(i,j)$). Therefore, we can define the joint influence strength of node under influence order ℓ.

Definition 6: *Node joint influence intensity. It is denoted that the node influence order is ℓ. Then, the joint influence intensity of* i *is expressed as follows.*

$$CI_\ell(i) = z_i \sum_{j \in Ball(i,\ell)} z_j \qquad (16)$$

In the work, the idea of greedy algorithm is used to better excavate the "super influence" node. It always removes the nodes with higher joint intensities and their edges in the network until the ratio of influence nodes reaches q_c. The specific algorithm is described as follows.

Step1: Initialize the influence node tag vector $(\boldsymbol{n} = (n_1, n_2, \ldots, n_N) = \boldsymbol{0})$. The influence order ℓ is given.

Step2: Remove the node $i = arg \max_{i \in N} CI_\ell(i)$ and its connected edges from the current network. It is denoted that $n_i = 0$.

Step3: It is determined whether the current matrix eigenvalue $\lambda_\ell^{current}(\boldsymbol{n})$ reaches the given threshold condition $\lambda(\boldsymbol{n}^*, q_c)$. If the condition is satisfied, then the algorithm stops; otherwise, it returns to Step 2. With the time complexity of $O(N \log N)$, CI algorithm removes the nodes of limited proportion. In influence analysis of large-scale complex social networks, CI algorithm can quickly search for influence nodes.

3   Experimental Analysis

The following two data sets are used in the work.

Table 1. Network parameters

Network name	Node number	Edge number	Network characteristics
Artificial network	50	245	Uniform network
Netscience network	1589	2742	Scale-free network

The above uniform network in the data set is an artificial network. The number of generated nodes is 50; the edge connection probability 0.2. As a real network, Netscience data set is used to describe cooperative relationship between scientists. Its degree distribution has obvious power-law characteristics. In the work, the node trust two-tuple was introduced into influence maximum node mining algorithm. Node trust measure is not the focus of the work. Therefore, uniformly distributed random numbers within $[0, 1]$ are generated by linear congruential generator. These numbers are used as the node initial trust and distrust of information in network propagation process. Table 1 shows the topology structure of two networks (See Figs. 3 and 4).

Fig. 3. Artificial uniform network

Fig. 4. Netscience network

Table 2. Eigenvalue thresholds and ratio of mining nodes in each network

Network type	Eigenvalues threshold	Ratio of nodes
1-level artificial uniform network	0.4548	0.6
2-level artificial uniform network	0.6755	0.6
3-level artificial uniform network	0.7696	0.6
1-level netscience network	0.0471	0.5544
2-level netscience network	0.2174	0.3612
3-level netscience network	0.3617	0.5538

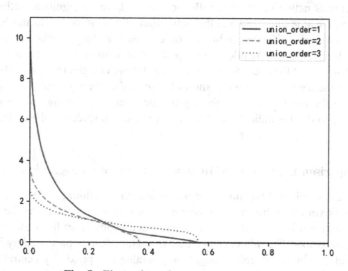

Fig. 5. Eigenvalues change curve (netscience)

3.1 Maximizing Influence Experiments

Section 2.3 describes the maximum node mining algorithm based on the percolation theory. Table 2 shows the node mining eigenvalue thresholds in different networks.

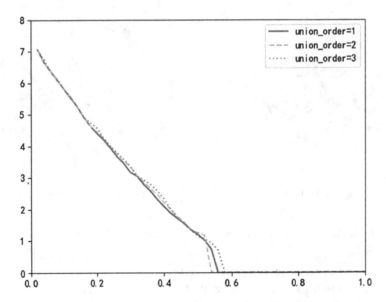

Fig. 6. Eigenvalues change curve (artificial uniform network)

Figures 5 and 6 show the change of maximum eigenvalue of weighted non-tracking matrix after removing the node with greatest joint influence intensity from the network. In homogeneous network, different influence orders have no significant effect on the change of eigenvalue threshold and the proportion of mining nodes. In the Netscience power-law network, different influence orders lead to large variation interval of eigenvalue threshold. Meanwhile, the proportions of mining nodes are also inconsistent. Data shows that the second-order influence mining can greatly reduce the number of mining influence nodes. Figure 5 shows the second-order eigenvalue change curve. After mining the node that meets the eigenvalue threshold, the eigenvalue decreases exponentially to 0. This indicates that the information can spread to the entire network faster.

3.2 Comparison Experiment of Influence Maximization Algorithm

In the work, we selected the influence maximization algorithm, node degree, eigenvector and proximity centrality based on percolation influence theory as the influence maximum node mining algorithm to mine influence nodes in artificial uniformity and Netscience networks. The influence intensity of each node was measured by indicators including joint influence intensity, degree, eigenvalue and proximity centrality.

The maximum influence nodes selected by different algorithms were used as seed nodes of linear threshold model. Then, the influence diffusion rate after model convergence was investigated to compare performances of different influence maximization algorithms. Figures 7 and 8 show the influence diffusion rates under Netscience and artificial uniformity networks. Results show that the degree of diffusion rate under uniformity network is as follows: improved percolation theory influence mining

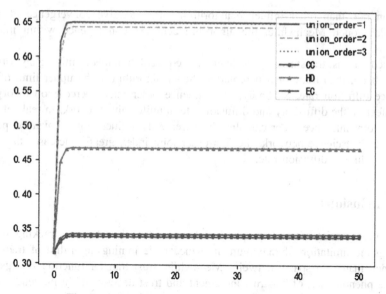

Fig. 7. Influence diffusion rate in netscience network

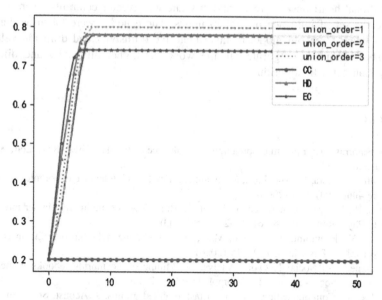

Fig. 8. Influence diffusion rate in artificial uniform network

algorithm ≥ traditional percolation theory influence node mining algorithm ≥ degree centrality ≥ eigenvector centrality ≥ proximity centrality.

In the uniformity network, the former four methods have similar diffusion utilities, which are significantly higher than that of proximity centrality. In Netscience power rate network, the influence mining algorithm based on percolation theory has higher diffusion rate than other influence algorithms after iterative convergence of linear threshold model. Meanwhile, the algorithm achieves convergence within minimal iterations.

Based on trust transmission mechanism, the percolation theory mining algorithm is slightly better than traditional percolation theory algorithm in the upper limit of final influence diffusion rate. Secondly, the influence order under percolation theory has little effect on the diffusivity and diffusion rate in uniformity network. In real network, suitable joint influence order can slightly increase the influence maximization performance. In Netscience network, the first-order joint index greatly increases the upper limit of influence diffusion rate.

4 Conclusions

In the work, social network graph model was established by non-tracking matrix, breaking the limitation of maximum influence node mining algorithm of traditional adjacency matrix in sparse network. Meanwhile, trust transfer function was used to describe phenomenon of "distrust increment and trust decrement" in influence propagation process. After mining influence node by percolation theory, we portrayed node influence by node joint influence propagation intensity index. The proposed algorithm and traditional heuristics (degree, proximity and eigenvector centrality) were used to excavate the set of influence maximum nodes. These nodes were taken as the seed nodes of linear threshold model to investigate the diffusivity and diffusion velocity. Result showed that the algorithm in the work has higher diffusivity and diffusion velocity than heuristic algorithm.

References

1. Pastorsatorras, R.: Epidemic spreading in scale-free networks. Phys. Rev. Lett. **86**(14), 3200–3203 (2001)
2. Crucitti, P., Latora, V., Marchiori, M., et al.: Error and attack tolerance of complex networks. Nature **406**(6794), 378 (2000)
3. Cohen, R., Erez, K., Ben-Avraham, D., et al.: Breakdown of the internet under intentional attack. Phys. Rev. Lett. **86**(16), 3682–3685 (2000)
4. Colizza, V., Flammini, A., Serrano, M.A., et al.: Detecting rich-club ordering in complex networks. Nat. Phys. **2**(3), 110–115 (2006)
5. Encyclopedia of Social Network Analysis & Mining. org.cambridge.ebooks.online.book. Author@dd. Soc. Netw. Anal. **22**(Suppl. 1),109–127 (2011)
6. Bavelas, A.: Communication patterns in task-oriented groups. J. Acoust. Soc. Am. **22**(6), 725–730 (1950)
7. Kitsak, M., Gallos, L.K., Havlin, S., et al.: Identification of influential spreaders in complex networks. Nature Phys. **6**(11), 888–893 (2010)
8. Kempe, D., Kleinberg, J., Tardos, É.: Maximizing the spread of influence through a social network. Progress. Res. 137–146 (2010)

9. Sehgal, U., Kaur, K., Kumar, P.: The anatomy of a large-scale hyper textual web search engine. In: International Conference on Computer & Electrical Engineering. IEEE Computer Society, pp. 491–495 (2009)
10. Granovetter, M.: Threshold models of collective behavior. Am. J. Sociol. **83**(6), 1420–1443 (1978)
11. Watts, D.J.: A simple model of global cascades on random networks. Proc. Nat. Acad. Sci. U.S.A. **99**(9), 5766–5771 (2002)
12. Chen, W., Wang, Y., Yang, S.: Efficient influence maximization in social networks. In: ACM SIGKDD International Conference on Knowledge Discovery and Data Mining, pp. 199–208. ACM (2009)
13. Hashimoto, K.I.: Zeta functions of finite graphs and representations of p-adic groups. In: Automorphic Forms & Geometry of Arithmetic Varieties, pp. 211–280 (1989)
14. Angel, O., Friedman, J., Hoory, S.: The non-backtracking spectrum of the universal cover of a graph. Trans. Am. Math. Soc. **367**(6), 4287–4318 (2015)
15. Klement, E.P., Mesiar, R., Pap, E.: Triangular norms. position paper I: basic analytical and algebraic properties. Fuzzy Sets Syst. **143**(1), 5–26 (2004)
16. Wu, J., Chiclana, F., Fujita, H., et al.: A visual interaction consensus model for social network group decision making with trust propagation. Knowl.-Based Syst. **122**(C), 39–50 (2017)
17. Morone, F., Makse, H.A.: Influence maximization in complex networks through optimal percolation. Nature **524**, 65–68 (2015)

Semi-supervised Classification of Concept Drift Data Stream Based on Local Component Replacement

Keke Qin[2] and Yimin Wen[1,2(✉)]

[1] Guangxi Key Laboratory of Trusted Software,
Guilin University of Electronic Technology, Guilin 541004, China
ymwen2004@aliyun.com
[2] School of Computer Science and Information Safety,
Guilin University of Electronic Technology, Guilin 541004, China
masterqkk@outlook.com

Abstract. Being compared with traditional data mining, data stream has three distinct characteristics which pose new challenges to machine learning and data mining. These challenges will become more serious when only few instances are labeled in data stream. In the paper, based on the algorithm of SPASC, a strategy of local component replacement for updating classifier pool is proposed. The proposed strategy defines a vector based on local accuracy to evaluate the adaptability of each "component" of a cluster-based classifier to a new chunk and makes the trained cluster-based classifiers in the pool adapt to the current concept better and faster while retaining as much learned knowledge as possible. The proposed algorithm is compared with the state of the art baseline methods on multiple datasets, the experimental results illustrate the effectiveness of the proposed algorithm.

Keywords: Semi-supervised learning · Local component replacement
Ensemble learning

1 Introduction

Being compared with traditional data mining, data stream has three distinct characteristics: i. the amount of data is very large, theoretically endless. ii. data generate very fast. iii. data are often accompanied with concept drift, i.e. the potential distribution of data over different time periods changes, IID no longer holds. This characteristic is extremely important in data streams. Concept drift specifically occurs in four types of gradual, incremental, abrupt and reappearance. These features of data stream pose new challenges to machine learning and data mining [1], including: (1) how to deal with concept drift; (2) how to update the trained models using new data.

However, the challenge will become more serious. In many real applications, labeling data is both laborious and costly, and the amount of data and the rapid data generation make it almost impossible to label data completely and timely. Therefore, it is more practical to handle data stream classification with few labeled data. So, a promising direction is semi-supervised learning [2, 3]. Semi-supervised classification

© Springer Nature Singapore Pte Ltd. 2018
Z.-H. Zhou et al. (Eds.): ICAI 2018, CCIS 888, pp. 98–112, 2018.
https://doi.org/10.1007/978-981-13-2122-1_8

of data stream is of great importance and has wide applications in many fields, such as network intrusion detection [4], software defect detection [5], etc. Many researchers [6–9] have conducted in-depth researches on semi-supervised classification of data stream with concept drift. Being compared with supervised environment, data stream classification in semi-supervised environment brings further new challenges: (1) how to construct classifier with instances that few are labeled; (2) how to detect concept drift under few labeled instances; (3) how to update history classifiers using the current data chunk that few are labeled.

The major contribution of Semi-Supervised Classification of concept drift data streams based on Local Component Replacement (SSCLCR) is to update the trained classifiers with a strategy of local component replacement which replaces some local components of a trained classifier with some local components of the newest trained classifier.

The rest of the paper is organized as follows: Sect. 2 is about the related works, and a top-level description of semi-supervised classification of data stream is introduced in Sect. 3. Section 4 is about the proposed SSCLCR for classification of data stream with concept drift, and the experimental evaluation of the proposed algorithm is introduced in Sect. 5. Finally, we conclude the paper in the last section.

2 Related Work

There has much research literatures [10, 11] of data stream classification in supervised environment and some surveys [1, 12, 13]. However, there are few studies on semi-supervised classification of data streams with concept drift, the related works can be roughly classified into two types: single model methods and ensemble methods.

2.1 Single Model Methods

Single model methods generally use one model to predict and incrementally update the trained model after its prediction to adapt to new concept. The related research works include [14–17].

Loo et al. [14] proposed a self-training style algorithm for online data stream classification algorithm, supervised K-means clustering is used to train base classifiers for initialization, each instance is predicted based on majority-class with the cluster who is nearest to it. The classifier is updated by merging high prediction confidence instances along with its predicted label (if unlabeled) or retrain using the labeled instances, and outdated and unused clusters are periodically erased. Zhu et al. [15] proposed an online Max-flow based algorithm, a weighted graph is constructed from the labeled and unlabeled instances and is later updated when new instances arrive. For classification tasks, minimized number of similar sample pairs are classified into different categories by calculating the min-cut over current max-flow, concept drift is handled through oMincut (online min-cut) incremental and decremental learning. Wu et al. [17] constructed an incremental decision tree using continuously arriving instances, a leaf is split if it receives a certain quantity of samples, concept clusters are periodically created at each leaf node of the tree through K-Modes clustering, the unlabeled instances are further labeled based on majority-class in training process, test instances are sequentially classified with the current tree in majority class. Concept drift is detected by measuring the deviation between the new concept clusters and the history concept clusters, if detected, corresponding leaf node is pruned. Li et al. [16] proposed REDLLA which is very similar with SUN [17] except for that REDLLA can detect recurring concept drift, which is based solely on distance between clusters.

2.2 Ensemble Methods

Ensemble learning methods [18] always handle data stream in chunks with fixed or variable size, base classifiers are trained from different chunks of the data stream by semi-supervised learning algorithm. These classifiers may represent different concepts that have been presented in the data stream and are combined in various ways for prediction [12].

Ahmadi et al. [19] proposed a self-training style ensemble algorithm SSEL, in which tri-training is used to train classifiers from both labeled and unlabeled instances in a chunk. Instances are classified using majority vote. Classifiers trained from the current chunk is used for incremental update. Feng et al. [20] proposed an incremental semi-supervised classification method IS3RS, which contains three phases, some representative instances are firstly selected by self-representative selection and then are predicted by KNN classifier under co-training framework, secondly high confidence instances together with the predicated label are added to labeled set to expand the number of labeled samples. Finally, classification is performed using KNN classifier based on enlarged labeled set.

In recent years, some researchers [6–9] constructed classifier based on semi-supervised clustering. Masud et al. [6] firstly designed classifier based on semi-supervised clustering. Impurity based K-means is proposed to construct classifier and label propagation is adopted to label the unlabeled cluster, each instance is classified based on inductive label propagation technique. Passive update is completed by eliminating the classifier with the worst performance on the current chunk after a chunk had been classified. Similarly, Hosseini et al. [7] proposed SPASC to construct classifier utilizing EM algorithm based on several assumptions, and classify the instances in a chunk sequentially with dynamic weight adjustment mechanism, a similarity based method is proposed for recurring concept drift detection, pool update is based on concept drift detection. Li proposed CASD [8] to create several sets of clusters by the weighted K-means to fit instance distribution. In classification process, if a instance is covered by a cluster model, the label of nearest cluster is predicted as its label, else, it's classified with another new classifier. Concept drift is detected if the time interval required for a given number of samples not to be covered is significantly reduced. Cluster model and base classifier are incrementally updated if concept drift is detected. Haque et al. [9] proposed a new framework named SAND, where classifier is created using impurity based K-means clustering and instance is classified by majority vote. CDT is proposed to estimate classifier's confidence. The oldest classifier is replaced with a new classifier trained based on confidence value if change is detected.

All the works above consider the semi-supervised scenarios where there are only a few instances are labeled in each chunk. Zhang et al. [21] considered a completely different case of semi-supervised scenario, where instances in a chunk maybe all labeled or all unlabeled. When a new chunk comes, instances in it are grouped into clusters, if the chunk is labeled, a classifier is also trained. A label propagation method that considers both class label from classifiers and clusters internal structure is utilized to infer class label of each cluster. All historical base models are weighted to form an ensemble for prediction and concept drift is implicitly alleviated through weight determination method.

3   Preliminary

3.1   Top Level Description

To begin with, data stream is assumed to be processed in chunks in which concept is single, denoted as $\{D_t\}_{t=1}^{\infty}$. In each chunk $\{D_t\}$, there are few labeled instances, denoted as (B_t, L_t), which are randomly labeled by "oracle", and more unlabeled instances, denoted as B_t'. When a new chunk D_t comes, instances in D_t are sequentially classified, then D_t is used to update the trained classifiers or classifier pool.

Next, we give a generic framework of data stream processing. Firstly, m classifiers are trained from the first m ($m = 1$ with default) chunks respectively, then these classifiers are added into the pool for initialization. Start from the $m + 1^{th}$ chunk, when a new data chunk comes, each instance in the chunk is classified consecutively. After the chunk has been classified, the concept in the chunk is checked. According to the result of concept detection, different strategies are selected to update the classifier pool. After it has been updated, repeat the above process on the next chunk. Many works are based on this mode [6, 7, 22]. Pseudo code 1 is a general algorithm template.

Pseudo code 1: Semi-supervised classification of data stream

Input: chunks of data: $\{(B_t, L_t) \cup B_t'\}_{t=1}^{\infty}$

Output: predict the label of each instance in each chunk

1 Initialization the classifier pool with the classifier trained on the first chunk;

2 **while true**

3      $(B_t, L_t) \cup B_t' = get\_next\_batch()$;

4      $Pool.\ classify\ (B_t, B_t')$;

5      $Pool.\ update\ (B_t, B_t', L_t)$;

6 **end while**

3.2   SPASC Algorithm

We firstly introduce the algorithm called semi-supervised pool and accuracy-based stream classification—SPASC [7], on which our algorithm is based. SPASC first initializes the pool with a classifier created using the first chunk, the weight of this classifier is set to 1, then iterates from the second chunk. When a new chunk comes,

SPASC selects the classifier with the max weight from the pool to classify each instance in the new chunk and each instance is predicted as the dominant class label of the cluster which is nearest to the it. After an instance is classified, if it has label, the weight of each classifier in the pool is tuned according to the individual classification result, else continues the next instance.

A recurring concept drift detection method is proposed in SPASC where whether recurring concept occurs or not is judged based on the similarity between each history classifier and the current chunk. Two methods of semi-Bayesian and semi-heuristic are used to measure the similarity.

In pool update phase, if recurring concept is not detected and the pool is not full, then a new classifier is trained on the current chunk and added into the pool, else incrementally update the most similar classifier. In the end, the weight of each classifier is recalculated based on the performance of it on the labeled instances of the current chunk.

Further, in the pool update phase, if the pool is full, the corresponding classifier is always incrementally updated whenever the concept from the current chunk belongs to a new concept or not, which leads to the updated classifier represents a hybrid knowledge, thus, the updated cluster-based classifier cannot perfectly adapt to the new environment. We expect to solve the above problem with a strategy of local component replacement, which can make the classifier adapt to new environment (concept) faster and better.

4   The Proposed Algorithm

The overall learning and classification flow of SSCLCR is illustrated in Fig. 1. We want to improve SPASC using the strategy of local component replacement. Based on different situations of the concept drift detection, we adopt different strategies to update the pool, specific ways to update the pool mainly include incremental update and local component replacement. The detail steps of semi-supervised method for updating classifier pool are shown in Pseudo code 2.

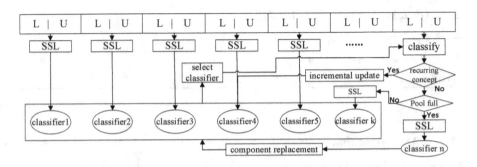

Fig. 1. The proposed algorithm architecture

4.1 Pool Update Strategy

Next, we give an in-depth introduction of incremental update and component replacement.

Incremental Update. In SPASC, if the concept of current chunk $\{X_i\}_{i=1}^b$ is a reappearance of a history concept or the pool is full, the classifier who is most similar current chunk in the pool is incrementally updated with the current chunk via EM algorithm. Being different with SPASC, in SSCLCR, incremental update component is only executed when a recurring concept drift is detected.

EM algorithm is utilized to determine hidden variable Z given known the new chunk of data $\{X_i\}_{i=1}^b$ and the old cluster centers ϕ_j ($1 \leq j \leq k$, k is number of cluster). Expectation and Maximization are two crucial steps of EM algorithm.

Expectation step: in this step, each instance X_i is assigned to a cluster according to probability $P(Z|X, \mu)$. Concretely, if X_i is unlabeled, then

$$P(Z_i = I_q | X_i, \mu) = \begin{cases} 1, & \text{if } \arg \min_{1 \leq j \leq k} \left\| X_i - \mu_{\phi_j} \right\| = q \\ 0, & \text{else} \end{cases} \tag{1}$$

where Z_i is a binary vector with k-dimensions, I_q is also a k-dimensions binary vector which only the qth element equals 1 while the other elements are all 0. So, $Z_i = I_q$ means $X_i \in \phi_q$, i.e. the ith instance belongs to the qth cluster, $1 \leq q \leq k$, μ_{ϕ_j} is center of the jth cluster ϕ_j and μ means all μ_{ϕ_j}. If X_i is labeled as l_i, then

$$P(Z_i = I_q | X_i, l_i, \mu) =$$
$$\begin{cases} 1, \text{if } \left(L \neq \varnothing \cap \arg \min_{j \in L} \left\| X_i - \mu_{\phi_j} \right\| = q \right) \text{or} \left(L = \varnothing \cap \arg \min_{1 \leq j \leq k} \left\| X_i - \mu_{\phi_j} \right\| = q \right) \\ 0, \text{else} \end{cases}$$
$$\tag{2}$$

where L represents the set of clusters index which the dominant label of a cluster is the same as that of instance X_i. L can be formulated as

$$L = \{ j | 1 \leq j \leq k, \arg \max_{1 \leq m \leq C} V_{\phi_j, m} = l_i \} \tag{3}$$

where $V_{\phi_j, m}$ is the number of instances in the cluster ϕ_j with the mth class label, C means the number of class label.

Maximization step: in this step, each cluster center is updated according to (4),

$$\mu'_{\phi_j} = \frac{1}{(N_{\phi_j} + N_j)} (\mu_{\phi_j} N_{\phi_j} + \sum_{i=1}^b Z_{i,j} X_i), \ 1 \leq j \leq k \tag{4}$$

where N_{ϕ_j} is the initial number of instance in ϕ_j, N_j is the number of instance in the current chunk $\{X_i\}_{i=1}^b$ which has been assigned to ϕ_j, $Z_{i,j}$ means the jth element of Z_i.

The Expectation step and Maximization step are iteratively executed until convergence or a stopping criterion is satisfied.

Incremental update is suitable when the difference between the current chunk and the classifier to be updated is small, so this strategy corresponds to the scenarios that the current data chunk is part of a reoccurring concept or a slight gradual or incremental drift of the existing concept.

Local Component Replacement. In SPASC, if the concept of the current chunk $\{X_i\}_{i=1}^b$ is a reappearance of a history concept or the pool is full, the classifier in the pool corresponding to the most similar concept is incrementally updated by semi-supervised EM algorithm [7]. However, being different with SPASC, when the pool is full and the current chunk is not a reappeared concept, instead of taking semi-supervised EM algorithm for incremental update as in SPASC, we take a strategy of local component replacement to update the most similar classifiers in the pool.

If the concept of the current chunk is a new concept and the pool is full, then a cluster-based classifier is created, which is denoted as $C_{new} = \{\phi_j\}_{j=1}^u$. Second, we evaluate the classification ability of each "component" ϕ_{ij} of the cluster based classifiers $\{C_i\}(C_i = \{\phi_{ij}\}_{j=1}^{k_i})$ in the pool that has made prediction for the current chunk, replace the poor performance "component" of C_i with the corresponding "component" of C_{new}.

For each cluster of C_i, we define a vector $V_i = \{V_{ij}\}_{j=1}^{k_i}$ to evaluate the adaptability of each "component" in C_i to the current chunk. V_{ij} can be defined as:

$$V_{ij} = \frac{NUC_{ij}}{NU_{ij}} \tag{5}$$

where NU_{ij} is used to count the number of instances which are classified by the "component" ϕ_{ij} of C_i, NUC_{ij} denotes the number of instances which are correctly classified by the "component" ϕ_{ij} of C_i.

Then replace the poor "component" of C_i with the nearest "component" of C_{new} to accomplish the update of the classifier C_i. The strategy of replacement is described as below,

$$\phi_{ij} = \begin{cases} \phi_{new,k}, \text{if } NU_{ij} \neq 0 \text{ and } V_{ij} < \tau \\ \quad\quad \text{and } \arg \min_{1 \leq m \leq u} \left\| \mu_{\phi_{ij}} - \mu_{\phi_{new,m}} \right\| = k \\ \phi_{ij}, \text{ else} \end{cases} \tag{6}$$

where μ_{ij} denote the centroid of the jth component of the ith classifier, τ is a pre-set constant which control the minimum adaptability. In this paper, τ is set to 0.5. Note that V is calculated and recorded in classification phase. Figure 2 is a simple diagram, assume classifier C_1 and C_2 participated in the classification of the instances of the current chunk, where poorly performed component $C_{1,3}$ of classifier C_1 is replaced with component $C_{new,3}$ of the new classifier C_{new} built from the current chunk, similarly, $C_{2,2}$ is replaced with $C_{new,2}$.

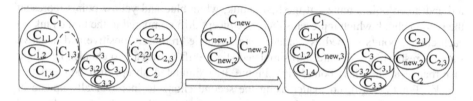

Fig. 2. Local component replacement diagram.

Pseudo code 2: Semi-supervised method for updating classifiers pool

Input: *Pool*, the classifier pool before updated, $(B_t, L_t) \cup B_t'$, the current chunk,

      maxC, pool size, *theta1*, the threshold for similarity

Output: *Pool*, the updated classifier pool

1 (max_*similarity*, *best_classifier*) = *Pool.assign_batch*(B_t, B_t', L_t);

2 **if** *max_similarity* > *theta1*

3   *best_classifier.incremental_update*(B_t, B_t', L_t);

4 **else**

5   C_{new} = *make_classifier*(B_t, B_t', L_t);

6   **if** *Pool.size*() = *maxC*

7   **for** *i* in 1 to *Pool.size*()

8   **for** *j* in 1 to *Pool[i].clusterNumber*

9   update the jth component of the ith classifier according to (6);

10   **end for**

11   **end for**

12   **else**

13      $Pool = Pool \cup \{C_{new}\}$;

14   **end if**

15 **end if**

16 **for** *i* in 1 to *Pool.size*()

17   *classifier_error* = *Pool[i].classify_error*(B_t, L_t);

18    $W_j = \beta^{2^{classifier\_error}}$;

19 **end for**

4.2   Analysis

To have a clear comparison between the strategies of incremental update and component replacement under different drift levels, we generate two chunk of data that their distributions are shown in Fig. 3, in which each sub-figure contains two chunks where red and green points represent the first chunk, while blue and yellow points represent the second chunk which represents there is concept drift compared to the first chunk. In each chunk, points shaped as "square" and "triangle" represent positive and negative samples respectively. Figure 3(a) simulated slight drift between two chunks, i.e. slight disturbance at the class boundaries. Figure 3(b) simulated abrupt drift between two chunks. For Fig. 3(a), no matter using the second chunk to incrementally update or conducting local component replacement to the first chunk, the results after update will not be much different. For Fig. 3(b), however, incrementally update the classifier trained from the first chunk with the second chunk of instances will causes some of the updated clusters to be very impure, as region III in Fig. 3(b), this problem is tackled with local component replacement. Local component replacement directly replace the outdated local component with a new component that reflect the distribution of the current data at corresponding position.

In short, it's necessary to adopt an appropriate pool update method based on different drift levels, local component replacement is more suitable than incremental update when abrupt concept drift occurs or there are server class overlap between two data chunk.

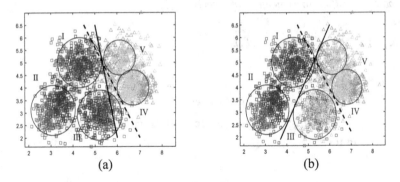

Fig. 3. Two chunks of data with different drift level. (Color figure online)

5   Evaluation

In this section, we discuss the datasets used in experiments, the experimental setup and the experimental results.

5.1   Experimental Data

We evaluate the proposed algorithm on 11 datasets, email list and spam data are derived from Spam Assassin Collection [23], for detail information, please refer to [24]. Spambase [25] is collected from postmaster. NEweather [26] is collected from the Offutt Air Force Base in Bellevue, Nebraska. Hyperplane [26] and Sea [26] are created using MOA [27], Hyperplane simulate incremental concept drift, 5% noise is introduced, Sea contains 4 concepts, concept drift is occurred according to f1 → f4 →

f3 → f2 → f3 → f4 → f3 → f2 → f1, 10% noise is introduced. Gaussian is a complex artificial dataset that features multiclass data, each drawn from a Gaussian distribution, with class means and variances changing according to the parametric equations described in [28]. Forest cover type [29] contains information of forest cover types for 30 × 30 m cells in four wilderness areas located in the Roosevelt National Forest of northern Colorado. Electricity [29] contains information of the price variation in the Australian New South Wales Electricity Market. Kddcup99 [7] is widely used in concept drift data stream field, a subset of 10% of all dataset is used in our experiments. Poker Hand [29] have instances representing all possible poker hands, each card in a hand is described by two attributes: suit and rank, the class indicates the value of a hand. Details of each dataset refer to Table 1. All datasets are normalized.

Table 1. Datasets

Dataset	#Feature	#Class	#Instance	Chunk size	Drift type
Email list	913	2	1500	100	abr/rec
Spam data	500	2	9324	100	gra
Spambase	57	2	4601	100	—
Hyperplane	10	2	200000	1000	inc
Sea	3	2	50000	500	abr/rec
NEweather	8	2	18159	360	—
Gaussian	2	4	6000	300	abr/gra
Forest cover type	54	7	581012	1000	—
Electricity	8	2	45312	1000	—
Kddcup99	41	23	494020	1000	—
Poker hand	10	10	829201	1000	—

[a.]*Abbreviation form. abr-abrupt, rec-reappearance, gra-gradual, inc-incremental*
[b.]

5.2 Experimental Setup

To evaluate the proposed algorithm, we compare it with two algorithms: SPASC [7], on which our work is based and ReaSC [6], another illustrious work in which classifier is cluster based. In SPASC, recurring concept drift detection is based on semi-supervised Bayes method. To make a fair comparison, according to the setting of relevant parameters in SPASC [7], pool size is set to 10, percentage of labeled instances in a chunk is 20%, cluster number is set to 5 for all algorithms. Especially for SSCLCR, parameter τ is set to 0.5. We implement the algorithm in Java environment with Weka [30] package, the experiments were run on Windows server 2012R2 with 64 GB of memory and 6 core CPU. Each chunk of instances is first used to test the current models and subsequently is used to update the models before the arrive of the next chunk. All experiments were run 10 times with differently labeled instances each time.

5.3   Experimental Results

Accuracy. Figure 4 shows accuracy evaluation of all three algorithms on 11 datasets. To have a more precise comparison, numerical results are presented in Table 2. It can be observed that SSCLCR is superior or at least equivalent to SPASC and ReaSC except for NEweather, Email list and Kddcup99, on which SSCLCR shows slightly lower accuracy. Especially, SSCLCR obviously outperform SPASC on Hyperplane, Forest cover type, Electricity, Poker hand and Gaussian datasets.

For Email data, abrupt concept drifts occur on the $4^{th}/7^{th}/10^{th}/13^{th}$ chunk, respectively, because the size of the pool is set to 10 and Email data set is too small, the strategy of local component replacement in SSCLCR begins to play a role after the 10th block, the effect is not particularly noticeable. From Fig. 4(a), however, it can observed that SSCLCR have faster performance improvement than SPASC on the 14th chunk.

For Sea dataset, abrupt concept drift is occurred every 10 chunks (on the $11^{th}/21^{th}/31^{th}/41^{th}/51^{th}/61^{th}/71^{th}/81^{th}/91^{th}$ block). It can be observed that SSCLCR always get higher accuracy than SPASC except on the 21th chunk.

For Gaussian dataset, abrupt concept drift occur on the 9th and 17th chunk, the centers of different class are also gradually moving regularly, the overlap of different classes is most serious on the 15th block. So, it can be observed that with respect to incremental update, local component replacement is a good way to reduce the impact of overlapping between different classes.

Table 2. Average accuracy (%) of 10 times run ($\pm$ the standard deviation of the accuracy) on each data streams with 20% instances labeled in each chunk. The values in boldface indicate the highest accuracy

Dataset	SSCLCR	SPASC	ReaSC
Spam data	**88.86 $\pm$ 0.75**	88.15 $\pm$ 1.08	81.76 $\pm$ 0.63
Hyperplane	**77.36 $\pm$ 0.82**	69.09 $\pm$ 0.97	50.49 $\pm$ 0.31
Sea	**81.00 $\pm$ 0.56**	78.86 $\pm$ 1.76	62.10 $\pm$ 0.74
NEweather	68.54 $\pm$ 0.86	**68.70 $\pm$ 1.02**	67.95 $\pm$ 0.00
Spambase	**97.65 $\pm$ 0.33**	97.52 $\pm$ 0.47	96.56 $\pm$ 0.14
Gaussian	**73.01 $\pm$ 2.00**	60.57 $\pm$ 0.90	32.89 $\pm$ 2.92
Cover type	**68.37 $\pm$ 0.64**	51.31 $\pm$ 5.52	66.24 $\pm$ 0.31
Electricity	**57.84 $\pm$ 0.63**	56.61 $\pm$ 1.26	56.71 $\pm$ 1.06
Email list	56.13 $\pm$ 3.01	**56.27 $\pm$ 2.71**	42.50 $\pm$ 5.99
Kddcup99	96.81 $\pm$ 0.37	**96.82 $\pm$ 1.12**	95.79 $\pm$ 0.10
Poker hand	**62.61 $\pm$ 0.28**	60.59 $\pm$ 1.36	53.53 $\pm$ 0.37

Considering ReaSC have no action for concept drift detection, the pool is regularly updated through discarding the worst performing classifier, while SPASC takes recurring concept drift detection, thus SPASC outperforms ReaSC on most of datasets (except for Forest cover type). However, being affected by the threshold parameter setting for recurring concept drift detection, concept drift detection result in SPASC is

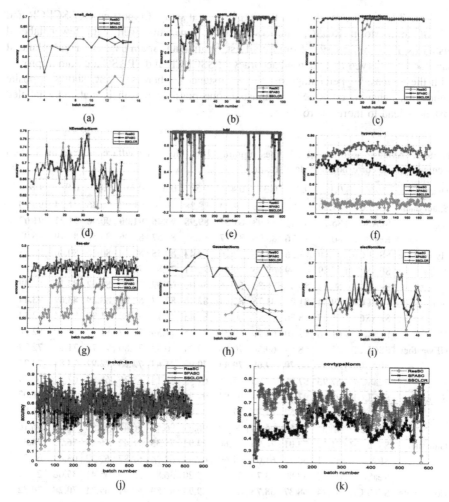

Fig. 4. Comparison of accuracy on each dataset with labeled percentage = 0.2

actually not very effective, resulting in more mistakes for concept drift detection, which maybe lead to a classifier is incrementally updated by a new chunk with different concept. SSCLCR alleviated this problem through local component replacement.

For SSCLCR, how many classifiers need to be updated and which component(s) of each classifier needed to be replaced could be automatically determined during the classification of the new chunk, so SSCLCR can update the pool more faster than SPASC does.

Effect of Percentage of Labeled Instances on Accuracy. To evaluate the effect of percentage of labeled instances on accuracy, we change the percentage of labeled instances from 10% to 100% with interval 10%.

Table 3 shows the results with this experimental setup. In most cases, SSCLCR and SPASC outperform ReaSC, and SSCLCR is especially well with fewer labeled instances, it can be observed from Email list, Spam data, Spam base and Hyperplane. It can also be observed that the total accuracy of SSCLCR and SPASC are both improved with the increase of percentage of labeled instances, whereas ReaSC shows opposite trend, this behavior on ReaSC is unusual, one possible reason is that more labeled instances lead to more micro clusters.

Table 3. Average accuracy of 10 times running. The values in boldface indicate the highest accuracy on the corresponding data stream.

Dataset		0.1	0.2	0.3	0.4	0.5	0.6	0.7	0.8	0.9	1.0
Spam data	SSCLCR	**87.20**	**88.86**	**89.48**	**89.90**	89.88	90.30	**91.07**	90.36	**91.24**	90.80
	SPASC	85.71	88.15	88.98	**89.37**	**89.98**	**90.46**	**91.00**	**90.96**	91.21	**91.07**
	ReaSC	80.40	81.76	83.06	83.77	84.33	84.57	85.06	85.52	85.79	86.02
Hyperplane	SSCLCR	**73.60**	**77.36**	**78.83**	**79.26**	**79.44**	**80.45**	**80.20**	80.26	80.40	82.83
	SPASC	65.59	69.09	69.35	70.48	70.71	71.85	72.52	73.37	74.06	75.54
	ReaSC	50.24	50.49	50.31	51.25	54.91	61.92	71.40	**81.02**	**87.14**	**88.49**
Sea	SSCLCR	**79.24**	**81.00**	**81.78**	**82.04**	**82.40**	**82.67**	**82.95**	**83.29**	**83.09**	**83.44**
	SPASC	76.63	78.86	80.47	80.72	81.43	81.74	81.69	81.87	82.34	82.14
	ReaSC	61.66	62.10	64.29	69.31	74.25	77.39	79.67	81.03	81.49	82.30
NEweather	SSCLCR	67.73	68.54	69.56	70.06	70.24	70.93	70.80	71.78	71.94	72.27
	SPASC	66.55	**68.70**	**69.69**	**70.40**	**70.99**	**71.63**	**72.23**	**72.97**	**73.81**	**73.75**
	ReaSC	**67.95**	67.95	67.95	67.95	67.95	68.14	68.76	68.83	67.81	66.59
Spambase	SSCLCR	**97.05**	**97.65**	**97.78**	**97.91**	**97.96**	**97.94**	**98.02**	**98.02**	**98.03**	**98.07**
	SPASC	96.42	97.52	97.77	**97.91**	97.94	**97.94**	**98.02**	**98.02**	**98.03**	**98.07**
	ReaSC	96.58	96.56	96.54	96.47	96.45	96.43	96.43	96.56	96.64	96.16
Gaussian	SSCLCR	**71.53**	**73.01**	**74.76**	**75.60**	**74.92**	**75.35**	**76.42**	**76.15**	**76.25**	**76.81**
	SPASC	59.21	60.57	61.07	61.60	62.33	62.34	62.87	62.52	63.63	65.21
	ReaSC	32.12	32.89	33.70	33..37	35.20	36.87	35.29	35.41	35.08	29.87
Cover type	SSCLCR	**67.74**	**68.37**	**68.73**	**69.29**	**69.93**	**69.89**	**69.91**	**69.71**	**70.28**	**70.32**
	SPASC	53.37	51.31	55.62	54.18	57.76	57.96	58.78	57.84	57.45	54.12
	ReaSC	66.02	66.23	66.03	65.78	65.80	66.02	66.12	65.92	66.13	66.08
Electricity	SSCLCR	**57.24**	**57.84**	**58.05**	**58.20**	**58.43**	**58.63**	**58.61**	**58.48**	**58.27**	**58.80**
	SPASC	56.48	56.61	56.53	56.56	57.19	57.33	57.41	57.18	57.19	57.93
	ReaSC	55.47	56.71	56.30	56.34	56.06	56.52	56.30	55.59	56.05	54.79
Email list	SSCLCR	**53.69**	56.13	**59.62**	62.32	64.81	66.81	**69.34**	**69.80**	71.97	74.64
	SPASC	**53.69**	**56.27**	59.44	**62.71**	**64.96**	**66.89**	69.29	69.73	**71.99**	**74.71**
	ReaSC	42.90	42.50	40.86	39.16	36.92	37.08	36.64	37.36	37..84	38.20
Kddcup99	SSCLCR	**97.29**	96.81	96.71	97.03	**97.22**	97.06	97.13	97.09	97.02	96.71
	SPASC	95.78	**96.82**	**96.76**	**97.07**	97.03	**97.08**	**97.16**	**97.27**	**97.42**	**97.42**
	ReaSC	95.76	95.79	95.77	95.64	95.81	95.83	95.89	95.93	95.96	94.23
Poker hand	SSCLCR	**61.94**	**62.61**	**62.84**	**62.88**	**62.88**	**62.86**	**62.97**	**62.74**	**62.46**	**61.83**
	SPASC	61.05	60.59	61.19	61.25	61.02	61.09	60.76	60.40	59.96	60.53
	ReaSC	53.82	53.53	53.19	52.87	52.79	52.47	52.34	51.92	51.95	54.74

6 Conclusion

This paper proposed an algorithm of SSCLCR based on SPASC. A novel strategy of local component replacement is utilized to update the pool with the new chunk. Experimental results verified the effectiveness of the proposed algorithm.

There are four research issues can be explored in the future: (1) Improve classification performance by incorporating transfer learning. (2) Try a more effective concept drift detection method in semi-supervised environment.

Acknowledgment. This work was partially supported by the National Natural Science Foundation of China (61363029, 61662014, 61763007), Guangxi Key Laboratory of Trusted Software (KX201721), Collaborative innovation center of cloud computing and big data (YD16E12), Image intelligent processing project of Key Laboratory Fund (GIIP201505).

References

1. Krawczyk, B., Minku, L.L., Woniak, M., Woniak, M., Woniak, M.: Ensemble learning for data stream analysis. Inf. Fusion **37**(C), 132–156 (2017)
2. Li, Y.F., Zhou, Z.H.: Improving semi-supervised support vector machines through unlabeled instances selection. In: Proceedings of the 25th AAAI Conference on Artificial Intelligence, pp. 386–391. AAAI, Menlo Park (2011)
3. Huang, K., Xu, Z., King, I., Lyu, MR.: Semi-supervised learning from general unlabeled data. In: Proceedings of the Eighth IEEE International Conference on Data Mining, pp. 273–282. IEEE, Piscataway (2008)
4. Breve, F., Zhao, L.: Semi-supervised learning with concept drift using particle dynamics applied to network intrusion detection data. In: Proceedings of BRICS Congress on Computational Intelligence & Brazilian Congress on Computational Intelligence, pp. 335–340. IEEE, New York (2013)
5. Zhang, Z.W., Jing, X.Y., Wang, T.J.: Label propagation based semi-supervised learning for software defect prediction. Autom. Softw. Eng. **24**(1), 1–23 (2016)
6. Masud, M.M., et al.: Facing the reality of data stream classification: coping with scarcity of labeled data. Knowl. Inf. Syst. **33**(1), 213–244 (2012)
7. Hosseini, M.J., Gholipour, A., Beigy, H.: An ensemble of cluster-based classifiers for semi-supervised classification of non-stationary data streams. Knowl. Inf. Syst. **46**(3), 567–597 (2016)
8. Li, N.: Clustering assumption based classification algorithm for stream data. Pattern Recog. Artif. Intell. **30**(1), 1–10 (2017)
9. Haque, A., Khan, L., Baron, M.: SAND: semi-supervised adaptive novel class detection and classification over data stream. In: Proceedings of Thirtieth AAAI Conference on Artificial Intelligence, pp. 335–340. AAAI, Menlo Park (2016)
10. Yang, L., Cheung, Y.M., Yuan, Y.T.: Dynamic weighted majority for incremental learning of imbalanced data streams with concept drift. In: Proceedings of the Twenty-Sixth International Joint Conference on Artificial Intelligence, pp. 2393–2399. AAAI, Menlo Park (2017)
11. Shao, J., Ahmadi, Z., Kramer, S.: Prototype-based learning on concept-drifting data streams. In: Proceedings of ACM SIGKDD International Conference on Knowledge Discovery and Data Mining, pp. 412–421. ACM, New York (2014)

12. Jayanthi, S., Karthikeyan, B.: A recap on data stream classification. Adv. Natural Appl. Sci. **8**(17), 76–82 (2014)
13. Wen, Y., Qiang, B., Fan, Z.: A survey of the classification of data streams with concept drift. CAAI Trans. Intell. Syst. **46**(11), 2656–2665 (2013). (In Chinese)
14. Loo, H.R., Marsono, M.N.: Online data stream classification with incremental semi-supervised learning. In: Proceedings of the Second ACM IKDD Conference on Data Sciences, pp. 132–133. ACM, New York (2015)
15. Zhu, L., Pang, S., Sarrafzadeh, A., Ban, T., Inoue, D.: Incremental and decremental max-flow for online semi-supervised learning. IEEE Trans. Knowl. Data Eng. **28**(8), 2115–2127 (2017)
16. Li, P.P., Wu, X.D., Hu, X.G.: Mining recurring concept drifts with limited labeled streaming data. ACM Trans. Intell. Syst. Technol. **3**(2), 1–32 (2012)
17. Wu, X.D., Li, P.P., Hu, X.G.: Learning from concept drifting data streams with unlabeled data. Neurocomputing **92**(9), 145–155 (2012)
18. Zhang, M.L., Zhou, Z.H.: Exploiting unlabeled data to enhance ensemble diversity. In: Proceedings of IEEE International Conference on Data Mining, pp. 619–628. IEEE, Piscataway (2010)
19. Ahmadi, Z., Beigy, H.: Semi-supervised ensemble learning of data streams in the presence of concept drift. In: Corchado, E., Snášel, V., Abraham, A., Woźniak, M., Graña, M., Cho, S.-B. (eds.) HAIS 2012. LNCS (LNAI), vol. 7209, pp. 526–537. Springer, Heidelberg (2012). https://doi.org/10.1007/978-3-642-28931-6_50
20. Feng, Z., Wang, M., Yang, S., Jiao, L.: Incremental semi-supervised classification of data streams via self-representative selection. Appl. Soft Comput. **47**, 389–394 (2016)
21. Zhang, P., Zhu, X., Tan, J., Guo, L.: Classifier and cluster ensembles for mining concept drifting data streams. In: Proceedings of IEEE International Conference on Data Mining, pp. 1175–1180. IEEE, Piscataway (2011)
22. Woolam, C., Masud, Mohammad M., Khan, L.: Lacking labels in the stream: classifying evolving stream data with few labels. In: Rauch, J., Raś, Zbigniew W., Berka, P., Elomaa, T. (eds.) ISMIS 2009. LNCS (LNAI), vol. 5722, pp. 552–562. Springer, Heidelberg (2009). https://doi.org/10.1007/978-3-642-04125-9_58
23. Apache Spam Assassin. http://spamassassin.apache.org
24. Machine Learning & Knowledge Discovery Group. http://mlkd.csd.auth.gr/concept_drift.html
25. Asuncion, A., Newman, D.J.: UCI Machine Learning Repository Irvine (2007)
26. Losing, V., Hammer, B., Wersing, H.: KNN classifier with self adjusting memory for heterogeneous concept drift. In: Proceedings of IEEE International Conference on Data Mining, pp. 291–300. IEEE, Piscataway (2017)
27. Bifet, A., Holmes, G., Kirkby, R., Pfahringer, B.: MOA: massive online analysis. J. Mach. Learn. Res. **11**(2), 1601–1604 (2010)
28. Elwell, R., Polikar, R.: Incremental learning of concept drift in nonstationary environments. IEEE Trans Neural Netw. **22**(10), 1517–1531 (2011)
29. Bifet, A., Pfahringer, B., Read, J., Holmes, G.: Efficient data stream classification via probabilistic adaptive windows. In: Proceedings of the 28th Annual ACM Symposium on Applied Computing, pp. 801–806. ACM, New York (2013)
30. Hall, M., Frank, E., Holmes, G., Pfahringer, B., Reutemann, P., Witten, L.H.: The WEKA data mining software: an update. ACM SIGKDD Explor. Newsl. **11**(1), 10–18 (2009)

Neural Networks and Deep Learning

A Fast and Accurate 3D Fine-Tuning Convolutional Neural Network for Alzheimer's Disease Diagnosis

Hao Tang[1(✉)], Erlin Yao[1], Guangming Tan[1], and Xiuhua Guo[2]

[1] Institute of Computing Technology, Chinese Academy of Sciences,
Beijing, China
tanghao@ncic.ac.cn
[2] Capital Medical University, Beijing, China

Abstract. The fast and accurate diagnosis of Alzheimer's Disease (AD) plays a significant part in patient care, especially at the early stage. The main difficulty lies in the three-class classification problem with AD, Mild Cognitive Impairment (MCI) and Normal Cohort (NC) subjects, due to the high similarity on brain patterns and image intensities between AD and MCI's Magnetic Resonance Imaging (MRI). So far, many studies have explored and applied various techniques, including static analysis methods and machine learning algorithms for Computer Aided Diagnosis (CAD) of AD. But there is still lack of a balance between the speed and accuracy of existing techniques, i.e., fast methods are not accurate while accurate algorithms are not fast enough. This paper proposes a new deep learning architecture to achieve the tradeoff between the speed and accuracy of AD diagnosis, which predicts three binary and one three-class classification in a unified architecture named 3D fine-tuning convolutional neural network (3D-FCNN). Experiments on the standard Alzheimer's disease Neuroimaging Initiative (ADNI) MRI dataset indicated that the proposed 3D-FCNN model is superior to conventional classifiers both in accuracy and robustness. In particular, the achieved binary classification accuracies are 96.81% and AUC of 0.98 for AD/NC, 88.43% and AUC of 0.91 for AD/MCI, 92.62% and AUC of 0.94 for MCI/NC. More importantly, the three-class classification for AD/MCI/NC achieves the accuracy of 91.32%, outperforming several state-of-the-art approaches.

Keywords: Alzheimer's disease · Deep learning · Brain MRI

1 Introduction

In 2010, the number of people over 60 years of age living with dementia was estimated at 35.6 million worldwide. This number is expected to almost double every twenty years [1]. Dementia refers to diseases that are characterized by loss of memory or other cognitive impairments and is caused by damage to nerve cells in the brain. Alzheimer's Disease (AD) is the most common form of dementia in elderly person worldwide, while Mild cognitive impairment (MCI) is a condition in which an individual has mild but noticeable changes in thinking abilities. Individuals with MCI are more likely to

© Springer Nature Singapore Pte Ltd. 2018
Z.-H. Zhou et al. (Eds.): ICAI 2018, CCIS 888, pp. 115–126, 2018.
https://doi.org/10.1007/978-981-13-2122-1_9

develop AD than individuals without [2]. Thus, effective and accurate diagnosis of AD especially for its early stage also known as MCI is required to reduce the costs related to care and living arrangements.

Early detection of the disease can be achieved by MRI, a medical imaging technique that uses strong magnetic fields, radio waves and filed gradients to generate images of the organs in the body. While the hazards of x-rays are now well-controlled in most medical contexts, MRI still may be seen as superior to CT in this regard. A multitude of existing pattern classification methods have been explored for this task in recent years, including support vector machines (SVM), multiscale fractal analysis and penalized regression [3, 4]. Some of these methods have proved a great performance in diagnosing AD from different types of neuroimaging data, sometimes even more accurate than experienced human radiologists [1]. Although conventional machine learning methods have yielded promising results, they still have several distinct drawbacks as follows: (1) Pipelines used in those traditional pattern classification methods mostly require multiple pre-processing steps for manual feature extraction. However, it would be a quite time-consuming, error-prone job, and the low-level descriptors fail to capture discriminative features, resulting in inferior classification results. (2) Since MRI creates a detailed 3D image of the brain, 3D neural networks play a significant part in aggregating information from all slices, while conventional methods using only single brain slices.

In this paper, we proposed an algorithm that used the whole 3D MRI blocks as input to train a 3D convolutional neural network and then finetuned model parameters on the validation data at the second stage. In the end, it is able to discriminate between diseased brains (AD) and healthy brains (NC) straightly. Moreover, we investigated several different neural network architectures and specifically chose to build 2D and 3D networks both similar to that of VGGNet [5] and ResNet [6] for image classification to demonstrate that 3D convolutions yielded better performance than 2D convolutions on slices in our experiments. We reported our classification results on the ADNI databases obtained using three binary classifiers (AD vs. NC, AD vs. MCI and MCI vs. NC) and an important 3-way classifier (AD vs. MCI vs. NC).

The reminder of this paper is organized as follows. In Sect. 2, we review related works on diverse classification methods that have been reported in the literature for this task. Section 3 describes the experimental data and MRI preprocessing steps briefly and focuses on introducing the deep learning approach. Section 4 addresses the experimental environment, parameter setup and performance evaluation of 2D and 3D convolutional neural network. Finally, the conclusion and future works are given in Sect. 5.

2 Related Works

Over the past few years, various of classification models have been tried for the discrimination of subjects using structural MRI and others modalities in Alzheimer disease. Among these approaches, traditional machine learning methods especially support vector machines (SVM) and its extended algorithms have been used widely in the area. [2] uses a whole brain and GM-based SVM for binary classification whose results are better than that achieved radiologists, which encourages and indicates a role for computerized diagnostic in clinical practices. [3] describes a kernel combination method for MRI, FDG-PET and CSF biomarkers which shows better performance compared to the case of using an individual modality. In [4], a multiscale fractal analysis approach is proposed to extract features of brain MRI, which is then classified by a SVM with polynomial kernel.

However, with the continuous development of deep learning methods, which has left an impressive result in many big data application fields such as image recognition, text mining and natural language processing in recent years, more and more researchers have similarly explored it for medical images analysis. In [7], an unsupervised method named AutoEncoder (AE) is first used to learn features from 2D patches extracted from either MRI scans or natural images. Moreover, the parameters learned by AE are applied and fixed as filters of a convolutional layer. Later on, [8] reports on a deep fully-connected network pre-trained with Stack Autoencoders (SAEs) and then classified with softmax regression for multi-class classification problem. While [9] implements an ensemble of deep learning architectures based Deep Belief Networks (DBN), another unsupervised learning algorithm consist of a stack of Restricted Boltzmann Machine (RBM) layers, and compares four different voting schemes for binary classification. It is worth mentioning that in [10], the algorithm of 2D and 3D convolutional networks combined with sparse AE respectively outperforms several other classifiers reported in the literature. And in a recent paper on ADNI data classification, the author of [11] proposed a Deeply Supervised Adaptive 3D-CNN (DSA-3D-CNN) which was pretrained by 3D-CAE on a source-domain for feature extraction and finetuning on a target-domain for classification. These two ideas are similar to ours and may perform better initial performance with some types of data, but both have a common restriction—high time complexity of weight initialization. In addition, other approaches either has complex operations to get hand-craft features for classification or has less experimental data to guarantee the robustness of the result, contrary to us. We demonstrate that our approach achieves relatively high accuracy and fast convergence, which is suitable for both binary and multi-class classification task without complex preprocessing or model stacking.

Table 1 summarizes several related studies mentioned above including the sample size and the reported performance. Although a direct comparison of these studies is difficult, as each study uses different datasets and preprocessing protocols, the table gives an indication of three typical measures—Accuracy (ACC), Sensitivity (SEN) and Specificity (SPE), achieved in the classification of MR images.

Table 1. Review of several methods for the AD, MCI and NC classification.

Paper	Sample size	ACC	SEN	SPE
[3]	AD 51, MCI 99, NC 52	AD vs. NC: 93.2%; MCI vs. NC: 76.4%	AD vs. NC: 93.0%; MCI vs. NC: 81.8%	AD vs. NC: 93.3%; MCI vs. NC: 66.0%
[8]	AD 65, cMCI 67, ncMCI 102, NC 77	AD vs. NC: 87.76% MCI vs. NC: 76.92% 4-way: 47.42%	AD vs. NC: 88.57% MCI vs. NC: 74.29% 4-way: 65.71%	AD vs. NC: 87.22% MCI vs. NC: 78.13% 4-way: 83.75%
[10]	755 patients from each class (AD, MCI and NC)	AD vs. NC: 95.39% MCI vs. NC: 86.84% AD vs. MCI: 92.11% 3-way: 89.47%	Null	Null
[11]	AD 70, MCI 70, NC 70	AD vs. NC: 99.3% MCI vs. NC: 94.2% AD vs. MCI: 100% 3-way: 94.8%	AD vs. NC: 100% MCI vs. NC: 97.1% AD vs. MCI: 100% 3-way: null	AD vs. NC: 98.6% MCI vs. NC: 91.4% AD vs. MCI: 100% 3-way: null

3   Materials and Methods

3.1   Experimental Data

The MRI data used in the preparation of this article were a subset obtained from the Alzheimer's Disease Neuroimaging Initiative (ADNI) database [12], a golden standard diagnostic approach with tests results and imaging comprising at least two years of follow-up. Since 2004, The ADNI has been validating the use of biomarkers including blood tests, tests of cerebrospinal fluid, and MRI/PET imaging for AD clinical trials and diagnosis, and now is in its fourth phase. ADNI is the result of many coinvestigators from a broad range of academic institutions and private corporations, and subjects have been recruited from over 50 sites across the U.S. and Canada. For our experiments the first part of the ADNI database consists of 3013 T1-weighted structural MRI (sMRI) scans from 321 subjects (112 NC, 129 MCI and 150 AD selected from the first two phases) is used to train a designed 3D-CNN, and then the optimal model parameters are loaded while finetuning on another part comprising 731 sMRI scans selected from the third phase. Figure 1 displays an example of three two-dimensional slices extracted from a MRI scan.

3.2   MR Image Preprocessing

The raw sMRI images from the ADNI database were provided in NII format (Neuroimaging Informatics Technology Initiative - NIfTI) for both three groups (AD, MCI, NC) with different resolution in both depth and dimension. First of all, sMRI images all have been spatially normalized by SPM12 [13] to guarantee that each image voxel corresponding to the same anatomical position. Next, all non-brain tissues, including

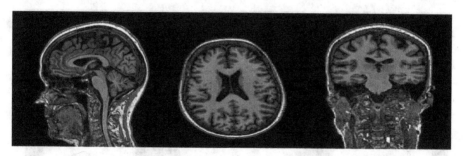

Fig. 1. Slices of a MRI scan of an AD patient. left: coronal plane, middle: axial plane, right: sagittal plane.

skull and neck voxels, were remove from the structural T1-weighted images using CAT12 [14], an extension toolbox for SPM12, by adopting a voxel-based morphometric (VBM) method [15]. In this step, a study specific grey matter template was then created using the VBM library and relevant protocol. After that, all brain-extracted images were segmented to grey matter (GM), white matter (WM) and cerebrospinal fluid (CSF). GM images were selected and registered to the GM ICBM-152 standard template using non-linear affine transformation. Furthermore, all 3D GM (one 3D image per subject) were averaged and resized to 92 * 96 * 96 voxels with voxel-sizes of 1.5 mm (Sagittal) * 1.5 mm (coronal) * 1.5 mm (axial), and concatenated into a stack (4D image = 3D images across subjects). Additionally, the modulated 4D image was then smoothed by a Gaussian kernel, sigma = 3 mm (standard sigma value in the field of MRI data analysis), which approximately resulted in full width at half maximums (FWHM) of 7 mm. As a result, the middle 62 slices are employed for the processing, whereas for lesion data, only slices that contain visual lesion features. Thus, each 3D dataset is divided into 62 * 96 * 96 boxes/blocks and be applied to train the 3D-FCNN model.

In parallel, in 2D form, all horizontal slices from preprocessed 3D GM are split apart as the training data of 2D-FCNN model, which means that each subject provides 62 slices labeled for the same class regardless of spatial information. As the following three-dimensional slices of a preprocessed sample with Gaussian smoothing, sigma = 3 mm are shown in Fig. 2.

3.3 Deep Neural Network

The proposed AD diagnostic model automatically extracts features of a brain MRI with a deep convolutional neural network and then finetunes on a task-specific classification subsequently. In this paper we also compare the performance of 2D and 3D convolutional network both on binary and 3-way classification.

2D Convolutions

Convolutional neural networks (LeCun 1998), also known as CNNs, are a specialized kind of neural network for processing data hierarchically that were inspired by the human visual system. CNNs have been tremendously successful in practical

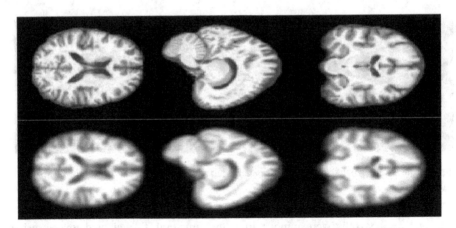

Fig. 2. Preprocessed image with Gaussian smoothing, sigma = 3 mm.

applications such as handwritten digit recognition [16], image segmentation [17] and object detection [18].

Equation (1) describes convolution operation in machine learning application. The input is usually a tensor of data and the kernel is usually a tensor of parameters that are adapted by the learning algorithm. Here we use a two-dimensional image I as our input, we probably also want to use a two-dimensional kernel K:

$$S(i,j) = (I * K)(i,j) = \sum_m \sum_n I(i+m, j+n)K(m,n) \qquad (1)$$

CNN leverages three main properties that can help improve a machine learning system: sparse interactions (also called local connectivity), parameter sharing and invariant representations. Moreover, convolution provides a means for working with inputs of variable size. The most general case (convolving over a zero padding input using non-unit strides) can be derived as Eq. (2).

$$o = \frac{i + 2p - k}{s} + 1 \qquad (2)$$

Where i, k, s, p devotes input size, kernel size, stride (distance between two consecutive positions of the kernel) and zero padding (number of zeros concatenated at the beginning and at the end of an axis), respectively. The o devotes output size, and the set of all these hidden units corresponds to a single "feature map".

A hidden layer has several feature maps and all the hidden units within a feature map share the same parameters. The parameter sharing feature is useful because it further reduces the number of parameters, and because hidden units within a feature map extract the same features at every position in the input. At a convolution layer, the previous layer's feature maps are convolved with learnable kernels and put through the activation function to form the output feature map. Each output map may combine convolutions with multiple input maps. In general, we have that:

$$x_j^l = f\left(\sum_{i \in M_j} x_i^{l-1} * k_{ij}^l + b_j^l\right) \tag{3}$$

Where M_j represents the input feature maps, f is the activation function, and $*$ denotes the convolution operation. In the computation the scalar term b_j^l is added to every entry of the array $x_i^{l-1} * k_{ij}^l$. Convolutional layers are followed by pooling layers. In a neural network, pooling layers provide invariance to small translations of the input. The most common kind of pooling is max pooling, which consists in splitting the input in (usually non-overlapping) patches and outputting the maximum value of each patch. Since pooling does not involve zero padding, the relationship describing the general case is just like Eq. (2) without $2p$.

3D Convolutional Neural Network (3D-CNN)
Similar to 2D convolution, the 3D convolution is achieved by convolving a 3D kernel to the cube (block) formed by stacking multiple consecutive slices together. In this architecture shown in Fig. 3, we consider using preprocessed sMRI scans with 62 slices of size 96 * 96 as a whole input block to the 3D-CNN model. We then apply a series of 3D convolutions with a kernel size of N * 3 * 3 (N in the temporal dimension and 3 * 3 in the spatial dimension) and p = 1 zero padding as well as 3D max-pooling with size of 2 * 2 * 2 on each of the 4 channels separately which resembles extensional VGG [5] model in space. In addition, we refer to [19] of adding two extra shortcut to merge low-level and high-level feature information, which is believed can alleviate gradient vanishing. Moreover, we adopt ReLU activation function [20], Batch Normalization [21], Dropout [22] and L2 regularization in order that accelerate the convergence of the model and prevent overfitting. The final outputs of size 280 * 3 * 6 * 6 are then flattened and used as inputs for a 3-layer fully connected neural network (i.e. with two hidden layers and an output layer). We choose two hidden layers with 1000 and 200 units with ReLU activation function and dropout layer, too. And an output layer with 3 units with a softmax activation function for 3-way classification (can also change into 2 units for binary classification). The 3 units in the output layer represent the posterior probability that the input belongs to each of the classes (AD, MCI and NC). Figure 3 provides an illustration of the network architecture.

We take the cross-entropy with a regularization term (also called a weight decay term) as our cost function $J(W,b)$, which is commonly used for multi-class classification tasks. This function is formulated as Eq. (4):

$$J(W,b) = -\frac{1}{N} \sum_{i=1}^{N} \sum_{j=1}^{C} I\{y^i = j\} \log\left(h_{W,b}(x^i)_j\right) + \lambda \sum_{l=1}^{n_l-1} \sum_{k=1}^{s_l} \sum_{m=1}^{s_{l+1}} (W_{mk}^l)^2 \tag{4}$$

Where N is the number of MRI scans, j is summing over the C classes (in this paper is equal to 2 or 3), $h_{W,b}$ is the score function computed by the network, x^i, y^i are the input and one-hot label of the ith scan as well as λ is weight decay parameter, respectively.

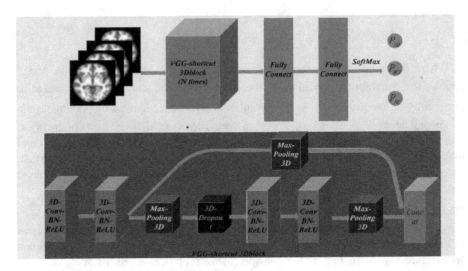

Fig. 3. Architecture of 3D-FCNN used for 3-way classification, where N = 3 in our experiment. As for binary classification, the output only has two units.

The 3D-CNN is trained by backpropagation (BP) algorithm with mini batch gradient descent. The weights of the hidden layer and output layer are randomly initialized. It is necessary to point out that since the model converged on the first part of training data, we would load the optimal weights of convolutional layer and frozen them while finetuning on another part. We also use Adam method [23] for first-order gradient-based optimization of stochastic objective functions, based on adaptive estimates of lower-order moments. In briefly, this method is computationally efficient and is well suited for problems that are large in terms of data and/or parameters as for our model.

4 Experiment and Performance Evaluation

We perform proposed model on the test data with a 10-fold validation strategy and obtain the average classification results of six binary classification and two 3-way classification tasks in Table 2. In order to keep in line with previous studies, we evaluate and compare the two approaches by using the same three metrics, that are accuracy (ACC), sensitivity (SEN) and specificity (SPE), detailed in Eq. (5).

$$ACC = \frac{TP + TN}{TP + TN + FP + FN} \qquad SEN = \frac{TP}{TP + FN} \qquad SPE = \frac{TN}{TN + FP} \qquad (5)$$

Where TP, TN, FP, FN denotes numbers of true positive, true negative, false positive and false negative classification results for a given set of data items, respectively. As we can observe from Tables 1 and 2, the 3D-FCNN model consistently outperforms several baseline methods and 2D-CNN on each performance measure.

Table 2. Results of proposed models on the testing dataset.

Classification	ACC (2D)	SEN (2D)	SPE (2D)	ACC (3D)	SEN (3D)	SPE (3D)
AD vs. NC	95.91%	95.1%	95.6%	96.81%	96.3%	97.2%
AD vs. MCI	86.84%	85.7%	87.6%	88.43%	88.0%	88.9%
MCI vs. NC	90.11%	88.9%	91.7%	92.62%	92.2%	93.1%
3-way	89.76%	Null	Null	91.32%	Null	Null

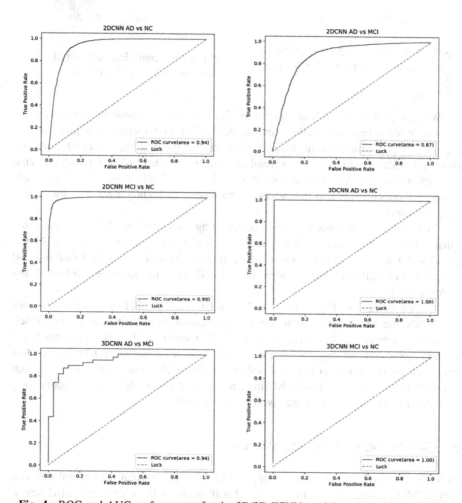

Fig. 4. ROC and AUC performances for the 2D/3D-FCNN model on the testing dataset.

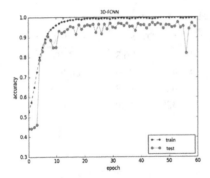

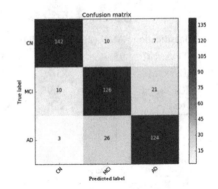

Fig. 5. The accuracy of 3-way classification on the training and test data over 50 epochs.

Fig. 6. The confusion matrix of 3D-FCNN classifier for 3-way classification.

We can see that 3D-CNN is superior to 2D convolution in these three measures. As we mentioned before, the results demonstrate that 3D model structure is more able to capture the spatial information, which is significant in medical images. In addition, our proposed model also outperforms several other methods listed in Table 1 with the same evaluation indexes. Although the method of [11] is slightly better than ours, their model is more complex and the process of parameter pretrained layer-wise by SAE is time-consuming.

Finally, Fig. 4 presents the receiver operating characteristic (ROC) of the three binary classifier, its performance is evaluated by the area under the ROC curve (AUC). Besides, the accuracy of the 3-way classification on the training and test data over 50 epochs as well as the confusion matrix of 3D-FCNN classifier are shown in Figs. 5 and 6, respectively.

As we can see from Fig. 4, for both AD vs. NC, AD vs. MCI and MCI vs. NC, our 3D-FCNN method achieves consistent improvement over those using 2D model. Moreover, compared with conventional methods mentioned in Sect. 2, our proposed model is more fast converge (~ 10 epochs) and accurate, as well as more robust on test data, with results shown in Fig. 5. Another significant observation from Fig. 6 is that our method can not only classify NC images well, but also achieve great classification accuracy for AD and MCI images with high similarity.

5   Conclusion and Future Works

In this paper, we have proposed a deep 3D fine-tuning convolutional neural network (3D-FCNN) architecture for the task of classification of brain structural MRI scans. Experimental results based on the ADNI dataset have demonstrated that 3D-FCNN can more accurately and effectively diagnose the AD than several state-of-the-art machine learning approaches listed in Table 1. Especially, we also compare the performance of 2D and 3D convolutions in a similar convolutional neural network architecture and hyper-parameters setting. The results indicate that 3DCNN has the potential to 'see' 3D context of MRI scans, e.g., amygdale, temporal pole and para-hippocampal regions, which are known to be related to AD by many studies [24, 25], whereas conventional 2D convolutional filters only capture 2D local patterns. Hence, the proposed 3D-FCNN outperforms traditional 2D convolutional networks.

Although the final classification accuracy of the proposed 3D-FCNN is a little lower than the approach of [11], our model balances the accuracy and speed without handcrafted feature generation and model stacking, which will better satisfy the clinical usage. In the future work, we will try more network architectures, e.g., the recently proposed CapsuleNet [26], which is said to have the equivariance to translation of object on images by using 'dynamic routing' algorithm rather than pooling after convolutions. This would mean a possibility of one-step analysis of complex MRI scans instead of multi-step pipelines such as skull stripping and re-alignment that currently dominate the filed. We believe that the system would be a very powerful tool for the future clinical diagnosis of Alzheimer's disease.

Acknowledgements. Data collection and sharing for this project was funded by the Alzheimer's Disease Neuroimaging Initiative (ADNI) (National Institutes of Health Grant U01 AG024904).

References

1. Prince, M., Bryce, R., Albanese, E., et al.: The global prevalence of dementia: a systematic review and metaanalysis. Alzheimer's Dement.: J. Alzheimer's Assoc. 9(1), 63–75 (2013)
2. Klöppel, S., Stonnington, C.M., Barnes, J., et al.: Accuracy of dementia diagnosis—a direct comparison between radiologists and a computerized method. Brain 131(11), 2969–2974 (2008)
3. Zhang, D., Wang, Y., Zhou, L., et al.: Multimodal classification of Alzheimer's disease and mild cognitive impairment. Neuroimage 55(3), 856 (2011)
4. Welch's, U.: Alzheimer's disease detection in brain magnetic resonance images using multiscale fractal analysis. ISRN Radiol. 2013(41), 627303 (2013)
5. Simonyan, K., Zisserman, A.: Very deep convolutional networks for large-scale image recognition. CoRR, vol. abs/1409.1556 (2014)
6. He, K., Zhang, X., Ren, S., et al.: Deep residual learning for image recognition. In: Computer Vision and Pattern Recognition, pp. 770–778. IEEE (2016)
7. Gupta, A., Maida, A.S., Ayhan, M.: Natural image bases to represent neuroimaging data. In: International Conference on Machine Learning, pp. 987–994 (2013)
8. Liu, S., Liu, S., Cai, W., et al.: Early diagnosis of Alzheimer's disease with deep learning. In: IEEE International Symposium on Biomedical Imaging, pp. 1015–1018. IEEE (2014)
9. Ortiz, A., Munilla, J., Gorriz, J.M., Ramírez, J.: Ensembles of deep learning architectures for the early diagnosis of Alzheimer's disease. Int. J. Neural Syst. 26(7), 1650025 (2016)
10. Payan, A., Montana, G.: Predicting Alzheimer's disease: a neuroimaging study with 3D convolutional neural networks. arXiv preprint arXiv:1502.02506 (2015)
11. Hosseiniasl, E., Ghazal, M., Mahmoud, A., et al.: Alzheimer's disease diagnostics by a 3D deeply supervised adaptable convolutional network. Front. Biosci. 23, 584–596 (2018)
12. http://www.loni.ucla.edu/ADNI
13. http://www.fil.ion.ucl.ac.uk/spm/software/spm12/
14. http://www.neuro.uni-jena.de/cat/index.html
15. Ashburner, J., Friston, K.J.: Unified segmentation. Neuroimage 26(3), 839–851 (2005)

16. Jarrett, K., Kavukcuoglu, K., Ranzato, M., LeCun, Y.: What is the best multi-stage architecture for object recognition? In: 2009 IEEE 12th International Conference on Computer Vision, pp. 2146–2153. IEEE (2009)
17. Long, J., Shelhamer, E., Darrell, T.: Fully convolutional networks for semantic segmentation. IEEE Trans. Pattern Anal. Mach. Intell. **39**(4), 640 (2017)
18. Ren, S., He, K., Girshick, R., et al.: Faster R-CNN: towards real-time object detection with region proposal networks. IEEE Trans. Pattern Anal. Mach. Intell. **39**(6), 1137–1149 (2015)
19. He, K., Zhang, X., Ren, S., et al.: Deep residual learning for image recognition. In: Proceedings of the IEEE Conference on Computer Vision and Pattern Recognition, pp. 770–778. IEEE (2016)
20. Nair, V., Hinton, G.E.: Rectified linear units improve restricted Boltzmann machines. In: International Conference on Machine Learning, pp. 807–814. Omnipress (2010)
21. Ioffe, S., Szegedy, C.: Batch normalization: accelerating deep network training by reducing internal covariate shift, 448–456 (2015)
22. Srivastava, N., Hinton, G., Krizhevsky, A., et al.: Dropout: a simple way to prevent neural networks from overfitting. J. Mach. Learn. Res. **15**(1), 1929–1958 (2014)
23. Kingma, D.P., Ba, J.: Adam: a method for stochastic optimization. arXiv preprint arXiv: 1412.6980 (2014)
24. Chetelat, G., Desgranges, B., De La Sayette, V., et al.: Mapping gray matter loss with voxel-based morphometry in mild cognitive impairment. NeuroReport **13**(15), 1939–1943 (2002)
25. Misra, C., Fan, Y., Davatzikos, C.: Baseline and longitudinal patterns of brain atrophy in MCI patients, and their use in prediction of short-term conversion to AD: results from ADNI. Neuroimage **44**(4), 1415–1422 (2009)
26. Sabour, S., Frosst, N., Hinton, G.E.: Dynamic routing between capsules. In: Advances in Neural Information Processing Systems, pp. 3859–3869 (2017)

Automatic Cloud Detection Based on Deep Learning from AVHRR Data

Meng Qiu, Haoyu Yin, Qiang Chen, and Yingjian Liu[✉]

Ocean University of China, Qingdao 266100, People's Republic of China
liuyj@ouc.edu.cn

Abstract. Thanks to the development of satellite remote sensing technology, more observing data are acquired and can be used for various purposes. However, statistical data show that half of the earth's surface is covered by clouds, which may seriously influence the usability of remote sensing data. Most existing cloud detection methods are manual or semi-automatic methods with low efficiency. This paper focuses on automatic cloud detection over sea surface from Advanced Very High Resolution Radiometer (AVHRR) data. A novel cloud detection framework named DBN-Otsu Hybrid Model (DOHM) has been proposed, which combines Deep Belief Networks (DBN) and Otsu's method for the first time. DOHM adopts adaptive thresholds to replace manual interventions, implementing full automation. Experimental results show that DOHM achieves the highest average accuracy ratio among the six detection methods. Moreover, DOHM makes a good balance between False Alarm Rate (FAR) and Miss Rate (MR).

Keywords: Deep learning · Cloud detection · Deep Belief Network
Advanced Very High Resolution Radiometer (AVHRR)

1 Introduction

With the development of satellite remote sensing technology, huge volumes of data are collected over large spatial-temporal scales with high resolution. However, statistical data show that half of the earth's surface is covered by clouds, which may seriously influence the usability of satellite remote sensing data. Cloud is an important climate factor. Cloud detection will help analyze weather conditions, environmental changes and climate characteristics. In addition, cloud detection and classification may contribute to prevent cloud related natural disasters, such as typhoon, torrential rain, lightning, etc., so as to reduce the loss of life and property. In some scientific researches, e.g. Sea Surface Temperature (SST) inversion from AVHRR data, the existence of cloud will seriously influence the availability of remote sensing data. Therefore, cloud detection has become the primary problem for remote sensing data applications.

Existing cloud detection methods can be roughly divided into two categories: traditional detection methods and new intelligent methods.

Z.-H. Zhou et al. (Eds.): ICAI 2018, CCIS 888, pp. 127–138, 2018.
https://doi.org/10.1007/978-981-13-2122-1_10

(1) Traditional Detection Methods

Traditional methods of cloud detection are mainly based on threshold segmentation. Threshold segmentation method has been evolving from the start and become increasingly mature. For example, Cazorla [1] uses the Cloud Mask method and contrasts with the CLAVR method. A two parts direct threshold technique and bi-spectral composite framework [2] also have been applied in cloud detection. In addition, hybrid threshold method [3] for cloud detection has caught many researcher's attention. Threshold method is evolving from fixed threshold to dynamic threshold [4], which can make algorithms have better flexibility, adaptability and accuracy [5]. However, with the continuous improvement of cloud detection technology, threshold methods have low degree of automation which is difficult to meet the needs of cloud detection.

Another kind of traditional detect system is cloud texture technology which is capable of extracting the texture features of the cloud [6] and other physical properties to distinguish cloud region [7]. Methods based on the singular value decomposition [8], SVM [9], bag-of-words model [10], Bayes [11] and progressive refinement scheme [12] are extensively used to extract the features of cloud. However, due to the improvement of remote sensing monitoring technology, cloud characteristics exceed the processing and analysis capacities of conventional cloud detection systems.

(2) New Intelligent Methods

After 90s, with the development of computer science and observation technology, machine learning technology ushered in the climax of the development [13]. This method attracts great attention due to its high accuracy, good adaptability, strong fault tolerance and high scalability. For example, Hinton, a pioneer in neural network learning, has made great achievements in the use of neural networks [14]. Although deep learning methods is widely used in face recognition and license plate recognition, a single deep learning method cannot meet the requirements of cloud detection because of the special characteristics of the cloud. Therefore, Larochelle [15] explores and optimizes the use of neural network method to train the classifier for cloud detection. In addition, some methods which integrate multiple detection methods [16] are welcomed by researchers since multiple methods can achieve better results [17]. For example, Xie [18] combines the deep learning and traditional neural network. Shi [19] integrates cloud mask technology and cloud segmentation algorithm, aiming at higher computing speed. These cloud detection methods are manual or semi-automatic, depending upon manual intervention with low efficiency. Automatic method enables higher detection efficiency. For example, Li et al. [20] call their method MFC which based on machine learning theory. In addition, CLOUDET algorithm combines Bias classifier and multilayer perceptron [21]. Fast cloud detection algorithm combining KNN classifier with cloud texture features [22]. These methods are automatic but their detection accuracy is not high. This paper proposes an automatic cloud detection method named DOHM, which combines DBN and Otsu's method for the first time, significantly improving the detection accuracy and efficiency. DOHM is simple to calculate, requires less variety of data, and can fully automate cloud detection.

The rest of this paper is organized as follows. Section 2 introduces the basic knowledge related to cloud detection, including the remote sensing characteristics of cloud, experimental data, and related methods. In Sect. 3, the framework of DOHM and the experimental steps for cloud detection are described in detail. In Sect. 4, experimental results of DOHM are compared with the other five methods in two aspects: visual analysis and quantitative analysis. Section 5 makes a conclusion.

2   Foundations of Cloud Detection

The cloud detection research involved in this article is mainly to analyze the remote sensing characteristics of the cloud, find the appropriate criteria and judgment methods for cloud detection, and thus perform cloud and sea surface classification. Cloud detection is the first step in satellite remote sensing data utilization and plays a crucial role in this entire process.

2.1   Remote Sensing Characteristics of Cloud

In remote sensing data obtained by multi-spectral scanning radiometer, the information at each pixel point is the radiance information of that point, showing the radiant energy per unit area of the pixel within a certain spectral region. With different compositional components, the reflectivity and other parameters in the spectral spectrum will be different. Specifically, the following three characteristics can be used to distinguish clouds and other substances.

(1) Contrast Signature. Contrast Signature is to distinguish different substances by using different reflection properties in solar spectral band (0.3–3 μm) and different infrared radiation in thermal infrared band (3–20 μm).
(2) Spectral Signature. Spectral signature shows there is a certain difference between the reflectivity of the same substance in different wavebands.
(3) Spatial Signature. Spatial signature mainly refers to the different laws of different substances in space.

2.2   Datasets

This paper uses AVHRR data in the experiment. The latitude and longitude ranges of experimental data are: 105.0°E–145.0°E, 10.0°N–50.0°N (see Fig. 1).

AVHRR is an advanced very-high-resolution radiometer on the NOAA series satellites with a spatial resolution of 1.1 km. AVHRR has evolved from four channels to five channels, including visible light red band, near infrared, mid-infrared, thermal infrared and other bands. In general, channel 1 and channel 2 are reflectance data, channel 3 to channel 5 are bright temperature data.

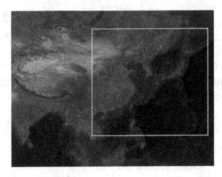

Fig. 1. The latitude and longitude ranges of experimental data. (105.0°E–145.0°E, 10.0°N–50.0°N)

2.3 Related Methods

DBN. DBN originates from machine learning and it was initially introduced by Hinton [23]. DBN is a probabilistic generative model, which is relative to the traditional discriminant model of neural network. Generative method offer a joint probability distribution based on data from observation and data label. DBNs are constituted by several layers of a kind of neural network named Restricted Boltzmann Machines (RBM). RBMs are formed by a single visible layer and an invisible layer. Links are established between the layers while units within a layer make no connection. Both hidden and visible layers exchange their information by generating weighs. Each visible and hidden node has two styles: active (value 1) and inactive (value 0). Here, the meaning of this sign is that computing model will choose nodes marked "1" other than nodes marked "0" into its computation.

In RBM, there is a weight w between any two connected neurons that represents the intensity of theirs connection. Each neuron itself has an offset coefficient b (for apparent neurons) and c (for hidden neurons) to express its own weight. Consider input vectors V and hidden layers output h, the energy of an RBM can be represented with Eq. (1).

$$E(V,h) = -\sum_{i=1}^{N_v} b_i v_i - \sum_{j=1}^{N_h} c_j h_j - \sum_{j=1}^{N_v, N_h} W_{ij} v_i h_j \qquad (1)$$

Otsu's Method. Otsu's method, proposed in 1979, is a significant method capable of making image segmentation using the maximum variance between the target and the background to determine the threshold of image segmentation dynamically. The Otsu method divides the target data into two parts, the background and the fore-ground, according to the gray characteristics of the target and the background. The greater the value of the variance, the more accurate the classification is. Otsu's method is one of the most classic nonparametric and unsupervised adaptive threshold selection algorithms.

3   Cloud Detection

3.1   DOHM

This paper proposes an algorithm framework named DOHM, which combines DBN and automatic threshold segmentation algorithm, i.e. Otsu's method. DOHM retains the advantages of DBN algorithm while achieving the automatic selection of thresholds.

The framework of DOHM is divided into two parts (see Fig. 2), DBN and Otsu. The artificially selected threshold portion of the DBN is replaced with Otsu, achieving automatic selection of threshold. In Fig. 2, "+1" is the offset of each layer, $h_i^{(j)}$ is the hidden layer, V_m' and $h_i^{(j)'}$ are the input recurrences, and $W_{pq}^{(n)}$ is the weight between the two nodes.

The input data of DOHM is a 9-dimensional feature vector $(v_1 - v_9)$: 5 channel data, the ratio of channel 2 and channel 1, the difference between channel 3 and channel 4, channel 3 and channel 5, Channel 4 and Channel 5. Steps of the DOHM algorithm are as follows:

(1)   Input feature vector $(v_1 - v_9)$. After data entry, network training is carried out. The final network structure is determined by adjusting parameters and fine-tuning. Take cloud detection as an example: The final optimal network structure has two layers, the first layer has six nodes, and the second layer has three nodes. The output of the first RBM in the training process is used as the input of the second RBM, each node sets the weight between the nodes, and the trained parameters will be passed to the corresponding NN network.

(2)   When the algorithm enters the last layer of the DBN, the Otsu is added, this paper uses Otsu instead of the max-method. Then the threshold is automatically determined by an adaptive method.

(3)   Get the final results.

3.2   Cloud Detection Experiment

Data Preprocessing. Preprocessing includes data cleaning, data tagging, and dataset establishment. Data cleaning is used to eliminate invalid data, data tagging is used to mark pixels. Each pixel in the marked data will have a label to indicate whether it is a cloud pixel. The establishment of the dataset should select the appropriate training set and test set according to the purpose of the experiment. In this paper, data of the 12 months of 2011 are set by the monthly labelling of 1–12 as the dataset used in the experiment. Dataset is divided into training set and test set: The training set is used for the training of DOHM model, and the test set is used to test the effect of network detection. Each dataset has a certain percentage of cloud data and clear data, which prevents the network from being heavily biased. In order to ensure the accuracy of the data, this experiment uses manual marking to make the label for each pixel.

Feature Extraction. Most traditional detection methods only select a certain channel or some channels as input, resulting in insufficient network training and deviation of

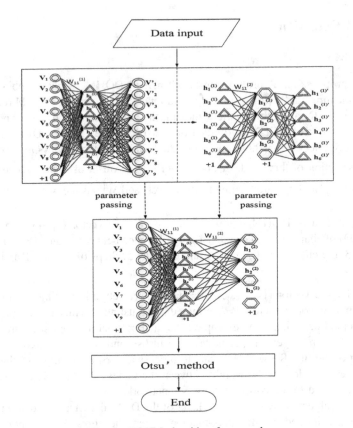

Fig. 2. DOHM algorithm framework.

detection results. To overcome this problem, DOHM selects all five channel data of AVHRR, and reconstructs the ratio of channel 2 and channel 1, the difference of channel 3 and channel 4, the difference of channel 3 and channel 5, and the difference of channel 4 and channel 5 as input of DOHM (as shown in Table 1).

Table 1. Vectors for experiment.

v_1	v_2	v_3	v_4	v_5	v_6	v_7	v_8	v_9
Ch1	Ch2	Ch3	Ch4	Ch5	$\frac{Ch2}{Ch1}$	Ch3–Ch4	Ch3–Ch5	Ch4–Ch5

The reflectance ratio between channel 2 and channel 1 is less than 0.75 in the clear sea surface and it is between 0.9 and 1.1 in the cloudy region. Given the methods of bi-spectral cloud detection, threshold of the difference of ch3 and ch4, ch3 and ch5, ch4 and ch5 can be roughly confirmed.

DOHM Training. G.E.Hinton proposed contrastive divergence capable of conducting pre-training in an unsupervised greedy layer-by-layer manner. The training process is to stack a certain number of RBM into a DBN. Data vector $(v_1 - v_9)$ and first hidden layer comprise the first RBM which has six hidden units and parameters will be generated of this layer after training (shown in Fig. 2). In this kind of network, the output of the previous RBM is the input of the next RBM so that the hidden layers from the first RBM can be transferred into the second RBM which has three hidden units.

The next step after pre-training is fine tuning. Fine tuning is supervised training process while pre-training is unsupervised. Network is organized by two hidden layers: the first layer is provided with 6 hidden layer units and the second layer is set 3 hidden layer units. Put the parameters into corresponding NN model of 9-6-3 network structure for supervised training after RBM's training. During the experimental process, DOHM adopts the maximum sample distance to detect the robustness of the network.

The key point of DOHM is taking the last layer of the DBN network as the input of the Otsu to determine the threshold dynamically. Otsu enables DOHM to achieve higher automation and strong adaptability relying on calculating threshold adaptively.

4 Experimental Results

In order to show the detection results of DOHM model more effectively and intuitively, this paper compares five other common detection methods with DOHM, including SVM, Probabilistic Neural Network (PNN), Neural Network (NN), naive Bayes classifier (Bayes), and DBN algorithm.

4.1 Visual Qualitative Analysis

In this paper, the land has been masked, and channel 4 is selected as the ground truth. The results of image visualization are shown in Figs. 3 and 4.

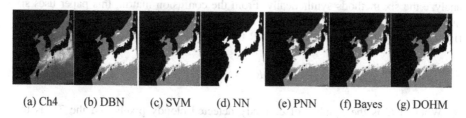

(a) Ch4   (b) DBN   (c) SVM   (d) NN   (e) PNN   (f) Bayes   (g) DOHM

Fig. 3. Results of six methods at 4:30 on May 5, 2011.

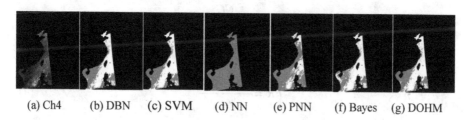

(a) Ch4   (b) DBN   (c) SVM   (d) NN   (e) PNN   (f) Bayes   (g) DOHM

Fig. 4. Results of six methods at 12:11 on December 2, 2011.

In order to demonstrate the pros and cons of these detection methods more comprehensively, images at different times within the same geographic location are selected for comparisons. In each figure, (a) is channel 4 image, (b) (c) (d) (e) (f) (g) are the experimental results of DBN, SVM, NN, PNN, Bayes, and DOHM. Obviously, DOHM, DBN and SVM perform much better than NN, PNN and Bayes. Further comparisons of the edges and details of cloud detection show that DOHM performs best and NN performs worst. It is noticed that all six methods performs less well in the winter and early spring seasons than in other seasons. The reason may be due to the snow and ice cover in high latitudes, showing the similar properties to the cloud at the bright temperature channel. However, DOHM still performs best among these six methods.

4.2   Quantitative Analysis

After visual comparisons of experimental results, further quantitative analysis is required. In order to test the accuracy and robustness of network classification, the 12 month datasets are divided into six groups as shown in Table 2. Each group choose two months with large differences in data, alternating training set and test set in two months. Twelve tests were carried out in total.

Table 2. Groups of datasets.

	Group.1	Group.2	Group.3	Group.4	Group.5	Group.6
Month 1	1	2	3	4	5	6
	7	8	9	10	11	12

Experiments choose 10000 pixels and use the confusion matrix (see Fig. 5) to analyze the six methods synthetically. From the confusion matrix, this paper uses six indicators: Recall Rate(RR), Precision Rate(PR), False Rate (FR), False Alarm Rate (FAR), Miss Rate(MR) and Ratio of RR to FR (RER) [24] (shown in Table 3).

$$RR = \frac{CC}{TC} \tag{2}$$

Where CC is the number of correctly detected cloudy pixels and the TC is the number of cloudy pixels over the sea surface.

$$PR = \frac{CC}{MC} \tag{3}$$

Where MC is the pixels detected as cloudy pixels

$$FR = \frac{CN + NC}{Ap} \tag{4}$$

Where *CN* is the number of cloudy pixels detected as clear pixels, *NC* is the number of clear pixels detected as cloudy pixels and the AP is the number of all pixels.

$$FAR = \frac{NC}{MC} \tag{5}$$

$$MR = \frac{CN}{TC} \tag{6}$$

$$RER = \frac{RR}{FR} \tag{7}$$

The RR and PR are reciprocal ones. In general, when RR is high, the PR will decrease. Since the cloud detection experiment focuses more on the accuracy of the detection, we are more concerned with the PR. PR shows that DBN and DOHM have the highest accuracy with relatively high RR. Although the PNN method has the highest RR, the PNN's PR is 74.47%, indicating that the accuracy of this method is very low. In addition, as far as the FR is concerned, DOHM minimizes the FR value, indicating that the DOHM method achieves the highest correct rate while ensuring a relatively high recall rate. In general, DOHM method can achieve a high degree of accuracy in the cloud detection problem.

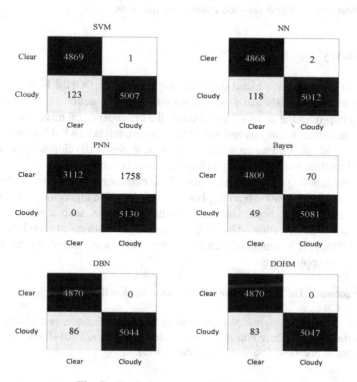

Fig. 5. Confusion matrix of six methods.

Table 3. Six indicators of different methods.

	SVM	NN	PNN	Bayes	DBN	DOHM
RR	97.60%	97.69%	**100%**	99.04%	98.32%	98.38%
PR	99.98%	99.96%	74.47%	98.64%	1	**1**
FR	1.24%	1.2%	17.58%	1.19%	0.86%	**0.83%**
FAR	0.02%	0.0399%	25.52%	1.359%	0	**0**
MR	2.40%	2.31%	**0**	0.96%	1.68%	1.61%
RER	78.71	81.41	5.688	83.23	114.32	**118.5**

FAR and MR are also a pair of contradictory indicators. Ideally, both indicators hope to be minimized. However, the practical resolution is to ensure that one indicator is below a certain threshold and let the other one to be as small as possible, so as to make a balance between them. In this paper, we are more concerned about FAR. DOHM and DBN perform best since their FAR is 0. Compared with MR of these two methods separately, the MR of DOHM is lower. Therefore, from the point of these two indicators, the DOHM method performs better. RER takes account of the relationship between RR and FR. The greater value means the better performance. The RER of DOHM is the largest among these six methods, so it has the best performance. Considering these six indicators, DOHM meets the requirements of cloud detection accuracy and balances the relationship between FAR and MR. In conclusion, DOHM performs best among these six cloud detection methods.

5   Summary

In this paper, an automatic cloud detection method named DOHM is proposed. This framework combines DBN and Otsu's methods to achieve automatic threshold selection, which improves the accuracy and reduces the dependence on manual intervention. From both quantitative analysis and visual qualitative analysis, DOHM achieves a good performance and has a broad application prospect. With the development of computer science and satellite remote sensing technologies, the requirements for automatic cloud detection are becoming more urgent. Future cloud detection methods tend to be diverse and comprehensive. Machine learning-based cloud detection algorithms will become a major trend in the future. Some latest deep learning networks, such as GoogleNet, may be applied in cloud detection. Although applying deep learning in cloud detection is still in its infancy, it is abundantly clear that this technology will shape the future of cloud detection applications.

Acknowledgement. The satellite data are provided by Satellite Ground Station of Ocean University of China.

This work is supported by the National Natural Science Foundation of China (61572448, 61673357), the Natural Science Foundation of Shandong Province (ZR2014JL043), and the Key R&D Program of Shandong Province (2018GSF120015).

References

1. Cazorla, A., Husillos, C., Antón, M., Alados-Arboledas, L.: Multi-exposure adaptive threshold technique for cloud detection with sky imagers. Sol. Energy **114**, 268–277 (2015)
2. Jedlovec, G.J., Haines, S.L., LaFontaine, F.J.: Spatial and temporal varying thresholds for cloud detection in GOES imagery. IEEE Trans. Geosci. Remote Sens. **46**(6), 1705–1717 (2008)
3. Andjunyang, Q.W.: A hybrid thresholding algorithm for cloud detection on ground-based color images. J. Atmos. Ocean. Technol. **28**, 1286–1296 (2011)
4. Liu, S., Zhang, Z., Xiao, B., Cao, X.: Ground-based cloud detection using automatic graph cut. IEEE Geosci. Remote Sens. Lett. **12**(6), 1342–1346 (2017)
5. Champion, N.: Automatic detection of clouds and shadows using high resolution satellite image time series. Int. Arch. Photogramm. Remote Sens. Spat. Inf. Sci. **XLI-B3**, 475–479 (2016)
6. Başeski, E., Cenaras, Ç.: Texture and color based cloud detection. In: International Conference on Recent Advances in Space Technologies, pp. 32–36. IEEE, London (2015)
7. Liu, S., Zhang, L., Zhang, Z., Wang, C., Xiao, B.: Automatic cloud detection for all-sky images using superpixel segmentation. IEEE Geosci. Remote Sens. Lett. **12**(2), 354–358 (2014)
8. Li, Z., Shen, H., Li, G., Xia, P., Gamba, L.Zhang: Multi-feature combined cloud and cloud shadow detection in GaoFen-1 wide field of view imagery. Remote Sens. Environ. **191**, 342–358 (2016)
9. Rossi, R., Basili, R., Frate, F.D., Luciani, M., Mesiano, F.: Techniques based on support vector machines for cloud detection on QuickBird satellite imagery. In: 2011 IEEE International Geoscience and Remote Sensing Symposium, pp. 515–518. IEEE, Vancouver (2011)
10. Yuan, Y., Hu, X.: Bag-of-words and object-based classification for cloud extraction from satellite imagery. IEEE J. Sel. Top. Appl. Earth Obs. Remote Sens. **8**(8), 4197–4205 (2015)
11. Xu, L., Wong, A., Clausi, D.A.: A novel Bayesian spatial-temporal random field model applied to cloud detection from remotely sensed imagery. IEEE Trans. Geosci. Remote Sens. **55**(9), 4913–4924 (2017)
12. Zhang, Q., Xiao, C.: Cloud detection of RGB color aerial photographs by progressive refinement scheme. IEEE Trans. Geosci. Remote Sens. **52**(11), 7264–7275 (2014)
13. Bengio, Y.: Learning deep architectures for AI. Found. Trends Mach. Learn. **2**(1), 1–127 (2009)
14. Hinton, G.E., Salakhutdinov, R.R.: Reducing the dimensionality of data with neural networks. Science **313**(5786), 504–507 (2006)
15. Larochelle, H., Bengio, Y.: Exploring strategies for training deep neural networks. J. Mach. Learn. Res. **10**, 1–40 (2009)
16. Zhu, Z., Wang, S., Woodcock, C.E.: Improvement and expansion of the Fmask algorithm: cloud, cloud shadow, and snow detection for Landsats 4–7, 8, and Sentinel 2 images. Remote Sens. Environ. **15**(9), 269–277 (2015)
17. Goff, M.L., Tourneret, J.Y., Wendt, H., Ortner, M., Spigai, M.: Distributed boosting for cloud detection. In: 2016 IEEE International Geoscience and Remote Sensing Symposium, pp. 134–146. IEEE, Beijing (2016)
18. Xie, F., Shi, M., Shi, Z., Yin, J., Zhao, D.: Multilevel cloud detection in remote sensing images based on deep learning. IEEE J. Sel. Top. Appl. Earth Obs. Remote Sens. **10**(8), 3631–3640 (2017)

19. Shi, C., Wang, Y., Wang, C., Xiao, B.: Ground-based cloud detection using graph model built upon superpixels. IEEE Geosci. Remote Sens. Lett. **9**(9), 1–5 (2017)
20. Li, Z., Shen, H., Li, H., Zhang, L.: Automatic cloud and cloud shadow detection in GF-1 WFV imagery using multiple features. In: 2016 IEEE International Geoscience and Remote Sensing Symposium (IGARSS), pp. 7612–7615. IEEE, Beijing (2016)
21. Islam, T., et al.: CLOUDET: a cloud detection and estimation algorithm for passive microwave imagers and sounders aided by Naïve Bayes classifier and multilayer perceptron. IEEE J. Sel. Top. Appl. Earth Obs. Remote Sens. **8**(9), 4296–4301 (2015)
22. Ufuk, D.U., Demirpolat, C., Demirci, M.F.: Fast cloud detection using low-frequency components of satellite imagery. In: Signal Processing and Communications Applications Conference. IEEE, Malatya (2017)
23. Hinton, G.E., Osindero, S., Teh, Y.W.: A fast learning algorithm for deep belief nets. Neural Comput. **18**(7), 1527–1554 (2006)
24. An, Z., Shi, Z.: Scene learning for cloud detection on remote-sensing images. IEEE J. Sel. Top. Appl. Earth Obs. Remote Sens. **8**(8), 4206–4222 (2015)

RBF Networks with Dynamic Barycenter Averaging Kernel for Time Series Classification

Hongyang Qin[1], Lifeng Shen[1], Chijun Sima[1], and Qianli Ma[1,2(✉)]

[1] School of Computer Science and Engineering,
South China University of Technology, Guangzhou 510006, China
scuterlifeng@foxmail.com, qianlima@scut.edu.cn
[2] Guangdong Key Laboratory of Big Data Analysis and Processing,
Guangzhou 510006, China

Abstract. Radial basis function (RBF) network has received many considerable realistic applications, due to its simple topological structure and strong capacity on function approximation. However, the core of RBF network is its static kernel function, which is based on the Euclidean distance and cannot be directly used for time series classification (TSC). In this paper, a new temporal kernel called Dynamic Barycenter Averaging Kernel (DBAK) is introduced into RBF network. Specifically, we first determine the DBAK's centers by combining k-means clustering with a faster DTW-based averaging algorithm called DTW Barycenter Averaging (DBA). And then, to allow the stable gradient-training process in the whole network, a normalization term is added to the kernel formulation. By integrating the warping path information between the input time series and the centers of kernel, DBAK based RBF network (DBAK-RBF) can efficiently work for TSC tasks. Experimental results demonstrate that DBAK-RBF can achieve the better performance than previous models on benchmark time series datasets.

Keywords: Radial basis function · Dynamic time warping
Kernel function · Time series classification

1 Introduction

Time series classification (TSC) is a hot research topic drawing wide attention in many fields, including atmospheric monitoring [17], clinical medicine [22], robotics [25], financial stock data analysis [23], etc. Recent years have witnessed a growing body of research works on time series classification (TSC). These research works can be roughly divided into three categories: (1) the first one is *distance*-based methods which aim to combine standard time series benchmark distance measures with other classifiers for TSC tasks. The mainstream approaches include 1NN-DTW model [4,9] combining one nearest neighbor (1NN) classifier with dynamic time warping (DTW) and its variant based

© Springer Nature Singapore Pte Ltd. 2018
Z.-H. Zhou et al. (Eds.): ICAI 2018, CCIS 888, pp. 139–152, 2018.
https://doi.org/10.1007/978-981-13-2122-1_11

on derivative distance called DD_{DTW} [10]; (2) the second one is *feature*-based classifiers. This category aims to extract representative features from raw time series and classify these time series by the learned discriminative features. The methods in this category include shapelet transform (ST) [12], learned shapelets (LS) [11], bag of SFA symbols (BOSS) [20], time series forest (TSF) [8], time series bag of features (TSBF) [3]; (3) the third category includes *ensemble* methods which present strong TSC baseline results. The two representative methods are Elastic Ensemble (EE) [14] and the collection of transformation ensembles (COTE) [1]. To be more specific, EE method is an ensemble classifier with 1NN based on 11 elastic distance measures, while COTE ensembles 35 different classifiers constructed in the time, frequency, change, and shapelet transformation domains. Although the above works all have made progress, TSC tasks remain to be challenging due to the fact that practical time series inevitably presents time-shift invariance, high-dimensionality and complex dynamics.

Radial basis function (RBF) network, first introduced by Broomhead and Lowe [5], is a simple and efficient tool for modeling time series. RBF networks typically have three layers: an input layer, a hidden layer with a non-linear RBF kernel function and a linear output layer. Given a D-dimensional sample x_i, a RBF network with P hidden neurons can be defined by the following equations:

$$\phi_{i,j} = \phi(\|x_i - c_j\|) = \exp\left(\frac{-\|x_i - c_j\|^2}{\sigma_j^2}\right) \tag{1}$$

$$y_{i,k} = f(\sum_{j=1}^{P} w_{j,k}\phi_{i,j} + b_k), \quad k = 1, 2, \cdots, K \tag{2}$$

where $\phi_{i,j}$ is the activation value of the j-th hidden neuron of this RBF network. $\phi(\cdot)$ denotes the RBF's kernel function and c_j is the center vector of the kernel function in the j-th hidden neuron. $y_{i,k}$ is the k-th output unit corresponding to the i-th input. f denotes the activation function. K is the number of output nodes. $w_{j,k}$ is the weight and b_k is the bias term.

Due to the existing of the RBF's kernel, RBF network is an universal approximator and can approximate any continuous function on a closed, bounded set with arbitrary precision when it has enough hidden neurons [15]. In this sense, RBF network is versatile, whose functions include function approximation, time series forecasting and nonlinear system identification, etc. However, RBF network has deficiencies in its kernel function as it is based on the Euclidean distance and cannot be directly used for time series classification. That is because the Euclidean distance assumes that each sample pair has been aligned with the same length, and time shifts or time distortions occur frequently and unpredictably in time series data.

In an attempt to adapt RBF kernel to time series classification, using DTW distance to replace the Euclidean distance is a feasible choice. The earliest work is the support vector machine (SVM) with a dynamic time-alignment kernel (DTAK) proposed by Shimodaira et al. [21]. DTAK uses inner product and kernel function to compute the distance between two aligned points (rather than

Euclidean distance) and then minimizes the accumulated distance like DTW. In a different manner, Bahlmann et al. directly replace the Euclidean distance with DTW and propose an invariant of the gaussian radius basis function called gaussian DTW (GDTW) kernel [2], while GDTW ignored the warping path information. Most recently, Xue et al. developed an altered Gaussian DTW (AGDTW) [24] kernel function to obtain corresponding transformed kernel features for time series classification. The main idea of AGDTW is to align time series at first and then calculate the values of the corresponding kernel function over the warping path. However, the AGDTW kernel is unnormalized and dependent on the length of warping path. Thus this kernel cannot be directly applied in gradient training-based RBF network.

In this paper, we propose a new temporal kernel called Dynamic Barycenter Averaging Kernel (DBAK) and combine it with RBF networks. DBAK is based on the AGDTW. Specifically, we first determine the DBAK's centers by k-means clustering with a DTW barycenter averaging (DBA) algorithm developed by Petitjean et al. [18]. Note that here the kernel center is a time series rather than a vector. Furthermore, to keep a stable gradient-training process in the whole network, a scaled term is added to normalize the proposed kernel. By integrating the warping path information between the input time series and the kernel's time series center, DBAK based RBF network can be efficiently applied in the TSC task.

The rest of this paper is organized as follows. In Sect. 2, we provide preliminaries related to our work. And then in Sect. 3, we describe the details of the proposed model DBAK-RBF. The experimental results are reported in Sect. 4. Finally, we make conclusions in Sect. 5.

2   Preliminaries

In this section, we will provide a brief review of dynamic time warping (DTW), AGDTW [24] and the dtw barycenter averaging (DBA) algorithm. More details about RBF network can be found in [5,15].

2.1   Dynamic Time Warping (DTW)

Dynamic time warping (DTW) [4] is a widely-used similarity measure for time series. It is based on the Levenshtein distance and aims to find the optimal alignment paths between two sequences. Given two sequences S and T defined by:

$$S = (a_1, a_2, \ldots, a_N) \tag{3}$$

$$T = (b_1, b_2, \ldots, b_M) \tag{4}$$

where N denotes the length of S and M denotes the length of T. And then, these two sequences can be arranged to an N-by-M grid $\mathcal{G}$, where the grid point $\mathcal{G}(i, j)$ denotes an alignment between the elements a_i from sequence S and elements b_j from sequence T. A warping path $W = (w_1, w_1, \ldots, w_K)$ between these two sequences is a sequence of points in grid $\mathcal{G}$ (K denotes the length of this path). This warping path satisfies three conditions:

1. boundary condition: this condition restricts endpoints of path by $w_1 = \mathcal{G}(1, 1)$ and $w_K = \mathcal{G}(N, M)$;
2. continuity: for each pair $w_k = \mathcal{G}(i_k, j_k)$ and $w_{k+1} = \mathcal{G}(i_{k+1}, j_{k+1})$, the warping path satisfies $i_{k+1} - i_k \leq 1$ and $j_{k+1} - j_k \leq 1$;
3. monotonicity: the points in the path must be monotonically ordered by time, e.g., $i_{k+1} - i_k \geq 0$ and $j_{k+1} - j_k \geq 0$;

In this way, searching for a DTW path is equivalent to minimizing all potential alignment paths in terms of cumulative distance cost. The DTW distance can be formally formulated by

$$DTW(S, T) = \min_W \sum_{k=1}^{K} \delta(w_k) \tag{5}$$

where δ denotes a distance between two aligned points a_{i_k} and b_{j_k}. In this work, we use the squared Euclidean distance:

$$\delta(a_{i_k}, b_{j_k}) = \|a_{i_k} - b_{j_k}\|_2^2 \tag{6}$$

Actually, the problem (5) can be solved by dynamic programming over a cumulative distance matrix Γ. The recurrence relation in DTW problem can be formulated by

$$\Gamma(i, j) = \delta(a_{i_k}, b_{j_k}) + \min\{\Gamma(i-1, j-1), \Gamma(i-1, j), \Gamma(i, j-1)\} \tag{7}$$

where $i = 1, 2, \ldots, N$ and $j = 1, 2, \ldots, M$. The initial conditions are

$$\Gamma(0, 0) = 0; \ \Gamma(i, 0) = \infty; \ \Gamma(0, j) = \infty \tag{8}$$

2.2   Altered Gaussian DTW (AGDTW)

Given two time series S and T defined in Eqs. 3 and 4, a GDTW kernel [2] can be formulated by

$$\mathcal{K}_{GDTW}(S, T) = \exp(-\frac{DTW(S, T)^2}{\sigma^2}) \tag{9}$$

where σ indicates the width of the kernel function and satisfies $\sigma \neq 0$. Although Eq. (9) is very similar to a general gaussian RBF kernel, it uses the DTW distance in the kernel rather than the Euclidean measure.

To integrate the warping path information in kernel function, an altered gaussian DTW (AGDTW) is developed. Let $\{a_{i_k}, b_{j_k}\}$ be the aligned point pair corresponding to the time series S and T, and the length of warping path is K, that is $k = 1, 2, \ldots, K$. Formally, AGDTW can be formulated by

$$\mathcal{K}_{AGDTW}(S, T) = \sum_{k=1}^{K} \exp(-\frac{\delta(a_{i_k}, b_{j_k})^2}{\sigma^2}) \tag{10}$$

2.3 DTW Barycenter Averaging (DBA)

DTW Barycenter Averaging (DBA) [18] is a global technique for averaging a set of time series. In our work, this technique will be used for estimating the kernel's centers. The main idea of DBA is to iteratively refine an initially (or temporary) average time series, in order to minimize its squared distance (DTW) to averaged sequences. Technically, the DBA works in two steps at each iteration:

1. Compute DTW distance between each time series sample and a temporary average sequence, and find associations between coordinates of the average sequence and coordinates of the set of time series;
2. Update each coordinate of the average sequence as the barycenter of coordinates associated to this coordinate at the first step;

More details about DBA can be found in [18].

In the next section, we will propose a new temporal kernel based on this AGDTW, which allows a stable gradient-training in the RBF networks.

3   DBAK-RBF Network

In this section, we explicate on our RBF network with a new dynamic barycenter averaging kernel (DBAK). We call our model DBAK-RBF for short.

3.1 Formulation

Assume that a time series dataset we used is $\mathcal{D} = \{T_1, T_2, \ldots, T_{N_D}\}$, which is a collection of N_D samples. Let a sample be $T_i = (x_1, x_2, \ldots, x_{L_{T_i}})$, where $x_n \in \mathcal{R}^d$ ($n = 1, 2, \ldots, L_{T_i}$), and L_{T_i} denotes the dimension of input time series. In this way, we can apply our DBAK-RBF in the time series classification on $\mathcal{D}$.

Our DBAK-RBF network is based on the topological structure of general RBF network and has a single hidden layer. The core of DBAK-RBF network is the DBAK function, which is formulated by

$$\mathcal{K}_{DBAK}(T_n, C_p) = \frac{\mathcal{M}}{K^2} \cdot \sum_{k=1}^{K} \exp\left(-\frac{\delta(x_{n,i_k}, c_{p,j_k})^2}{\sigma_p^2}\right) \tag{11}$$

where C_p denotes the center of p-th hidden neuron's activation function. $\mathcal{M} = \max\{L_{T_n}, L_{C_p}\}$. It is worth mentioning that in the case of the general RBF network, the center is a vector; while in our DBKA-RBF network, the center is a time series. The reason behind this selection will be explained later. The parameter σ denotes the width of the kernel function. K is the length of the warping path between time series T_n and C_p. L_{T_n} and L_{C_p} are the length of T_n and C_p, respectively. Compared with AGDTW kernel in Eq. 10, our DBKA has a normalization term in Eq. 11, which can be decomposed into two parts:

$$\frac{\mathcal{M}}{K^2} = \frac{1}{K} \cdot \frac{\mathcal{M}}{K} \tag{12}$$

Here, the part $\frac{1}{K}$ is used to eliminate the effects of the warping length K. In AGDTW, values of kernel easily turn out to be much larger due to a large K, which will hinder the gradient-training. The $\frac{M}{K}$ is the other scaling term and can make the outputs of $\mathcal{K}_{DBAK}(T_n, C_p)$ tend to a standard normal distribution, as is shown in Fig. 1. Its effectiveness will be verified in later experiments.

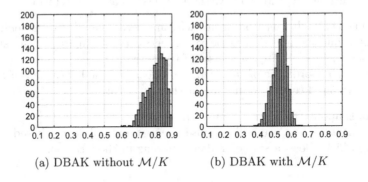

(a) DBAK without $\mathcal{M}/K$ (b) DBAK with $\mathcal{M}/K$

Fig. 1. Effects of the scaling term $\mathcal{M}/K$

After computing the DBAK kernel's values in the hidden layer, we can formulate the c-th activation value $z_{n,c}$ as following

$$z_{n,c} = \sum_{p=1}^{P} w_{p,c} \cdot \mathcal{K}_{DBAK}(T_n, C_p) + b_c \tag{13}$$

where $c = 1, 2, \cdots, C$ and $p = 1, 2, \cdots, P$. Using the **Softmax** function, we can have the posterior distribution $y_{n,c}$ of the c-th class corresponding to the n-th time series sample:

$$y_{n,c} = \textbf{Softmax}(z_{n,c}) = \frac{\exp(z_{nc})}{\sum_{\hat{c}=1}^{C} \exp(z_{n\hat{c}})} \tag{14}$$

In this way, the classification target function can be defined by the well-known cross entropy loss:

$$\mathcal{L}_{\mathcal{D}} = -\frac{1}{N_D} \sum_{n=1}^{N_D} \sum_{c=1}^{C} \hat{y}_{n,c} \log y_{n,c} \tag{15}$$

where $\hat{y}_n$ denotes the true label of the n-th time series.

3.2   Training

The whole process of training our DBAK-RBF includes two steps: (1) the selection of the centers $\{C_p\}$ in Eq. 11 and (2) the estimation of other parameters, including the width $\{\sigma_p\}$, the weights $\{w_{p,c}\}$ and the biases $\{b_c\}$.

Algorithm 1. DBAK-RBF

Require: $\mathcal{D} = \{T_1, \cdots, T_{N_D}\}$: a set of time series; $\mathcal{Y} = \{y_1, \cdots, y_{N_D}\}$ corresponding labels of $\mathcal{D}$; C: the number of kernel's centers.

1. Clustering the given data $\mathcal{D}$ by k-means;
2. Using DBA algorithm to obtain the centeral time series in each cluster;
3. Initialize parameters $\{\sigma_p\}$, $\{w_{p,c}\}$, and $\{b_c\}$;
4. Forward-propagation by using Eq. 11, Eq. 13, Eq. 14 and Eq. 15;
5. Back-propagation by using gradients formulated by Eq. 16-19;

return the centers $\{C_p\}$; the optimized width $\{\sigma_p\}$, the weights $\{w_{p,c}\}$ and the biases $\{b_c\}$.

Step 1: Selection of Centers in DBAK. For traditional RBF networks, the kernel's centers $\{C_p\}$ usually are optimized by clustering algorithm like k-means. However, these techniques cannot be directly used in our DBKA-RBF network since they depend on the Euclidean distance metric and thus are not suitable for time series data. Using Dynamic Time Warping (DTW) is a natural way to select the centers when modeling time series data. However, this change will generate much higher computation costs because we need to compute the DTW distances of all sample pairs. Here, we combine the DTW-based averaging method called DTW Barycenter Averaging (DBA) [18] with k-means clustering to estimate our DBAK function's center time series $\{C_p\}$. Firstly, we use the k-means method to roughly divide the training set into different clusters. And then, for each cluster, DBA algorithm is used to learn an averaged time series as a center of DBAK function. This strategy is also suggested by the work [16].

Step 2: Gradient-Based Optimization. After fixing the centers in DBAK function, we can use gradient-based optimization techniques to learn other parameters, including $\{\sigma_p\}$, $\{w_{p,c}\}$ and $\{b_c\}$. According to Eqs. 11 to 15, we can easily obtain the corresponding gradient formulations.

The gradient of the width σ of DBAK function is given by

$$\frac{\partial \mathcal{L}_\mathcal{D}}{\sigma_p} = \frac{1}{N_D} \sum_{n=1}^{N_D} \sum_{c=1}^{C} (y_{n,c} - \hat{y}_{n,c}) w_{p,c} \cdot \frac{\partial \mathcal{K}_{DBAK}(T_n, C_p)}{\sigma_p} \tag{16}$$

$$\frac{\partial \mathcal{K}_{DBAK}(T_n, C_p)}{\sigma_p} = \frac{2\mathcal{M}}{\sigma_p^3 K^2} \sum_{k=1}^{K} \exp\left(-\frac{\delta(x_{n,i_k}, c_{p,j_k})^2}{\sigma_p^2}\right) \delta(x_{n,i_k}, c_{p,j_k})^2 \tag{17}$$

where $\mathcal{M} = \max\{L_{T_n}, L_{C_p}\}$ and p denotes the width of p-th hidden neuron's activation function.

As for the weights $\{w_{p,c}\}$ and the biases $\{b_c\}$, we have

$$\frac{\partial \mathcal{L}_D}{w_{p,c}} = \frac{1}{N_D} \sum_{n=1}^{N_D} (y_{n,c} - \hat{y}_{n,c}) \mathcal{K}_{DBAK}(T_n, C_p) \tag{18}$$

$$\frac{\partial \mathcal{L}_D}{b_c} = \frac{1}{N_D} \sum_{n=1}^{N_D} (y_{n,c} - \hat{y}_{n,c}) \tag{19}$$

With these gradients, we use the well-known ADAM method to do the stochastic optimization. More details about ADAM can be found in [13]. Algorithm 1 details the complete process of training our DBAK-RBF.

3.3 Discussion of Computational Complexity

In this section, we discuss the computational complexity of our model. The main computational cost in our model is calculation of the DTW distances. There are two parts where we use DTW. The first one is to compute center vectors with DBA. Here we need to compute the DTW distance between each time series sample and a temporary average sequence in an iterative way. The second is to compute the warping path between the given input time series and the centroids in the hidden neurons.

The original DTW algorithm has a quadratic time and space complexity. To accelerate the computation of DTW, FastDTW is an efficient and effective DTW variant, which can provide a linear time and space complexity [19]. For simplicity, given two time series with same length T, and then the worst-case behavior of FastDTW can be given by $\mathcal{O}(T \times (8r + 14)) \approx \mathcal{O}(T)$, where r denotes the limit radius in FastDTW (a small positive constant value).

In this way, the computational complexity in the first part can be given by $\mathcal{O}(mTN_{iter})$, which is linear to the number of time series samples m and the length of time series T. N_{iter} denotes the number of iteration. The complexity of the second part is $O(nT)$ where n is the number of hidden neurons, whose complexity is also linear to the length T. According to these analysis, we can conclude that our model do not bring much more computational cost and can be efficiently used in linear time.

4 Experiments

In this section, we will thoroughly evaluate the performance of our DBAK-RBF by comparing it with other mainstream TSC methods.

The classification accuracy is used as performance measures and can be defined by

$$\text{Accuracy} = \frac{\text{the number of correctly classified data}}{\text{the total number of testing data}} \tag{20}$$

4.1   Comparisons Analysis

To examine the performance of the proposed DBAK-RBF, we perform a series of experiments on a publicly available Time Series datasets from the "UCR Time Series Data Mining Archive" [6]. UCR datasets exhibit different background knowledge in a wide range of application domains, which is very effective in testing the learning and classification ability of time series classifiers.

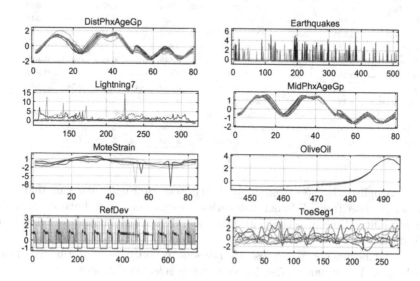

Fig. 2. Visualization of Six UCR Datasets

Table 1. The details of 15 selected UCR datasets

DATASETS	♯ CLASSES	♯ TRAIN	♯ TEST	LENGTH
DistalPhalanxOutlineAgeGroup	3	139	400	80
DistalPhalanxOutlineCorrect	2	276	600	80
DistalPhalanxTW	6	139	400	80
Earthquakes	2	139	322	512
Lightning2	2	60	61	637
Lightning7	7	70	73	319
MiddlePhalanxOutlineAgeGroup	3	154	400	80
MiddlePhalanxTW	6	154	399	80
MoteStrain	2	20	1252	84
OliveOil	4	30	30	570
ProximalPhalanxOutlineAgeGroup	3	400	205	80
ProximalPhalanxTW	6	205	400	80
RefrigerationDevices	3	375	375	720
ToeSegmentation1	2	40	228	277
Wine	2	57	54	234

Table 2. Accuracy of 9 mainstream TSC classifiers and our DBAK-RBF

DATASETS	DTW$_{1NN}$	DD$_{DTW}$	ST	LS	BOSS	TSF	TSBF	EE	COTE	DBAK-RBF
DistPhxAgeGp	0.770	0.705	0.770	0.719	0.748	0.748	0.712	0.691	0.748	**0.855**
DistPhxCorr	0.717	0.732	0.775	0.779	0.728	0.772	0.783	0.728	0.761	**0.845**
DistPhxTW	0.590	0.612	0.662	0.626	0.676	0.669	0.676	0.648	0.698	**0.800**
Earthquakes	0.719	0.705	0.741	0.741	0.748	0.748	0.748	0.741	0.748	**0.835**
Lightning2	0.869	0.869	0.738	0.820	0.836	0.803	0.738	**0.885**	0.869	0.869
Lightning7	0.726	0.671	0.726	0.795	0.685	0.753	0.726	0.767	0.808	**0.863**
MidPhxAgeGp	0.500	0.539	0.643	0.571	0.546	0.578	0.578	0.558	0.636	**0.805**
MidPhxTW	0.507	0.487	0.520	0.507	0.546	0.565	0.597	0.513	0.571	**0.647**
MoteStrain	0.835	0.833	0.897	0.883	0.879	0.869	0.903	0.883	**0.937**	0.890
OliveOil	0.833	0.833	0.900	0.167	0.867	0.867	0.833	0.867	0.900	**0.933**
ProxPhxAgeGp	0.805	0.800	0.844	0.834	0.834	0.849	0.849	0.805	0.854	**0.878**
ProxPhxTW	0.761	0.766	0.805	0.776	0.800	0.815	0.810	0.766	0.781	**0.825**
RefDev	0.464	0.445	0.581	0.515	0.499	**0.589**	0.472	0.437	0.547	**0.589**
ToeSegmentation1	0.772	0.807	0.965	0.934	0.939	0.741	0.781	0.829	**0.974**	0.943
Wine	0.574	0.574	0.796	0.500	0.741	0.630	0.611	0.574	0.648	**0.907**
Avg. Rank	8.067	8.533	4.367	6.533	5.667	5.000	5.167	6.900	3.233	1.533
Difference	6.533	7.000	2.833	5.000	4.133	3.467	3.633	5.367	1.700	-

We visualize the spectrogram of 6 datasets in Fig. 2. We can find that time series data from different categories present time-shift invariance, high-dimensionality (time direction), and complex dynamics. Thus, classifying time series data is not a trivial task.

In our experiments, we choose 15 datasets to verify the performance of our model. Table 1 summarizes the details of these selected datasets.

As for the compared baselines, we select 9 mainstream representative TSC models. They come from three categories we introduced in Sect. 1, including (1) distance-based methods; (2) feature-based methods and (3) ensemble-based methods.

In more details, for the distance-based methods, we compare with DTW$_{1NN}$ [9] and DD$_{DTW}$ [10]. For the feature-based methods, we select shapelet transform (ST) [12], learned shapelets (LS) [11], bag of SFA symbols (BOSS) [20], time series forest (TSF) [8] and time series bag of features (TSBF) [3]. For the ensemble-based methods, Elastic Ensemble (EE) [14] and the collection of transformation ensembles (COTE) [1] are chosen.

We report our comparison results in Table 2. As shown in this table, DBAK-RBF outperforms other TSC models with the average rank of 1.533 and achieves the best performance on 12 of the 15 datasets. The ensemble-based COTE and the feature-based ST model achieve the second and the third best performance, respectively. On the whole, ensemble-based and feature-based methods are better than the distance-based methods. Notably, our DBAK-RBF achieves much higher accuracy on some TSC tasks. For example, on the MidPhxAge dataset, DBAK-RBF can achieve an accuracy of 0.805, while the second best one ST only has an accuracy of 0.643. This is a huge improvement. For Wine, our DBAK-RBF works very well and achieves an accuracy of 0.907, while other classifiers achieve accuracies of 0.6481 by $COTE$, 0.574 by DTW_{1NN}, DD_{DTW}, and EE. Moreover, there are also some datasets like Lightning2 on which feature-based methods do not perform well. On the Lightning2, DBAK-RBF has an accuracy of 0.8689, but feature-based methods have accuracies of 0.7377 (ST), 0.8197 (LS), 0.8361 ($BOSS$), 0.8033 (TSF) and 0.7377 ($TSFB$), respectively.

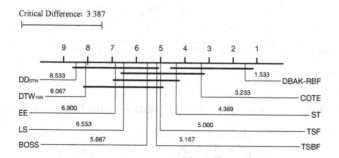

Fig. 3. Comparison of DBAK-RBF with other 9 classifiers on the Nemenyi test. Groups of classifiers that are not significantly different ($p = 0.05$) are connected.

In addition, we conduct non-parametric tests (Friedman test, Nemenyi post-hoc test) to make statistical comparisons [7]. The statistic τ_F of Friedman test is 15.08, which is larger than the critical value 1.950 ($p = 0.05$). Thus, the null-hypothesis (all the algorithms are equivalent) is rejected and we can proceed with a post-hoc Nemenyi test. The critical difference diagram compares the results of DBAK-RBF to other classification methods with the Nemenyi test, which is shown in Fig. 3. The performance of two classifiers is significantly different if the corresponding average ranks differ by at least a critical difference (CD) value. As seen in Fig. 3, the CD value here is 3.387. From this result, we can conclude that although our DBAK-RBF is only slightly better than COTE and ST, it significantly outperforms most of the other models.

4.2   Components Analysis

To further verify the effectiveness of the components (e.g. DBAK, normalization term) in our model, we conduct the components analysis in this section. We firstly compare DBAK-RBF with two baseline models on UCR datasets.

The first model is a regular RBF network. In this case, we allow this network to be directly applied in TSC task by regarding an input time series as an L_{T_i}-dimension input vector. As seen in Fig. 4a, we verify that our DBAK is better than the Euclidean distance (ED) based original RBF kernel (as there are more blue points in the figure).

In Fig. 4b, we use the k-means based on ED to fix centers in our DBAK-RBF, without using the DBA. Again, the result shows that our DBAK is much better than the case of using ED-based k-means.

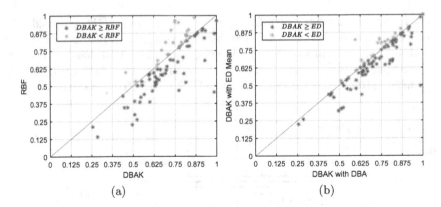

Fig. 4. (a) DBAK-RBF vs. RBF; (b) DBAK with DBA/ED-mean

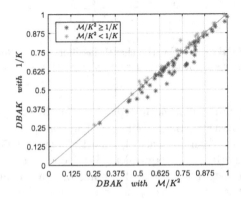

Fig. 5. DBAK with $\mathcal{M}/K^2$ vs. DBAK with $1/K$

Furthermore, we also investigate the performance of our normalization term $\mathcal{M}/K^2$ in Eq. 11. We compare our DBAK with $\mathcal{M}/K^2$ and the case with $1/K$ in Fig. 5. This result shows that the scaling term $\mathcal{M}/K$ significantly improves the performance on time series classification.

5   Conclusion

In this work, we develop a new DTW-based kernel, named Dynamic barycenter averaging kernel (DBAK), in RBF network to address the problem of time series classification. The main characteristics of the DBAK-based RBF network include: (1) we use a two-step k-means clustering based on a DTW-Barycenter-Averaging (DBA) algorithm to estimate the centers of RBF network's kernel function. (2) we add a normalization term to the kernel formulation to allow a stable gradient-training process. Experiment results demonstrate that our DBAK-RBF has a better or more competitive TSC performance than previous

TSC methods, including some strong baselines, such as ensemble-based COTE and feature-based ST.

Acknowledgment. The work described in this paper was partially funded by the National Natural Science Foundation of China (Grant No. 61502174), the Natural Science Foundation of Guangdong Province (Grant No. 2017A030313355, 2017A030313358, 2015A030313215), the Science and Technology Planning Project of Guangdong Province (Grant No. 2016A040403046), the Guangzhou Science and Technology Planning Project (Grant No. 201704030051), the Opening Project of Guangdong Province Key Laboratory of Big Data Analysis and Processing (Grant No. 2017014), and the Fundamental Research Funds for the Central Universities (Grant No. D2153950).

References

1. Bagnall, A., Lines, J., Hills, J., Bostrom, A.: Time-series classification with cote: the collective of transformation-based ensembles. IEEE Trans. Knowl. Data Eng. **27**(9), 2522–2535 (2015)
2. Bahlmann, C., Haasdonk, B., Burkhardt, H.: Online handwriting recognition with support vector machines - a Kernel approach. In: Proceedings Eighth International Workshop on Frontiers in Handwriting Recognition, pp. 49–54 (2002)
3. Baydogan, M.G., Runger, G., Tuv, E.: A bag-of-features framework to classify time series. IEEE Trans. Pattern Anal. Mach. Intell. **35**(11), 2796–2802 (2013)
4. Berndt, D.J.: Using dynamic time warping to find patterns in time series. In: KDD Workshop, pp. 359–370 (1994)
5. Broomhead, D.S., Lowe, D.: Multivariable functional interpolation and adaptive networks. Complex Syst. **2**(3), 321–355 (1988)
6. Chen, Y., et al.: The UCR time series classification archive, July 2015
7. Demsar, J.: Statistical comparisons of classifiers over multiple data sets. J. Mach. Learn. Res. **7**, 1–30 (2006)
8. Deng, H., Runger, G., Tuv, E., Vladimir, M.: A time series forest for classification and feature extraction. Inf. Sci. **239**, 142–153 (2013)
9. Ding, H., Trajcevski, G., Scheuermann, P., Wang, X., Keogh, E.: Querying and mining of time series data: experimental comparison of representations and distance measures. Proc. VLDB Endow. **1**(2), 1542–1552 (2008)
10. Górecki, T., Luczak, M.: Using derivatives in time series classification. Data Min. Knowl. Discov. **26**(2), 310–331 (2013)
11. Grabocka, J., Schilling, N., Wistuba, M., Schmidt-Thieme, L.: Learning time-series shapelets. In: Proceedings of the 20th ACM SIGKDD International Conference on Knowledge Discovery and Data Mining, KDD 2014, pp. 392–401. ACM, New York (2014)
12. Hills, J., Lines, J., Baranauskas, E., Mapp, J., Bagnall, A.: Classification of time series by shapelet transformation. Data Min. Knowl. Discov. **28**(4), 851–881 (2014)
13. Kingma, D.P., Ba, J.: Adam: a method for stochastic optimization. CoRR abs/1412.6980 (2014)
14. Lines, J., Bagnall, A.: Time series classification with ensembles of elastic distance measures. Data Min. Knowl. Disc. **29**(3), 565–592 (2015)
15. Park, J., Sandberg, I.W.: Universal approximation using radial-basis-function networks. Neural Comput. **3**(2), 246–257 (1991)

16. Petitjean, F., Forestier, G., Webb, G.I., Nicholson, A.E., Chen, Y., Keogh, E.: Faster and more accurate classification of time series by exploiting a novel dynamic time warping averaging algorithm. Knowl. Inf. Syst. **47**(1), 1–26 (2016)
17. Petitjean, F., Inglada, J., Gançarski, P.: Satellite image time series analysis under time warping. IEEE Trans. Geosci. Remote Sens. **50**(8), 3081–3095 (2012)
18. Petitjean, F., Ketterlin, A., Ganarski, P.: A global averaging method for dynamic time warping, with applications to clustering. Pattern Recogn. **44**(3), 678–693 (2011)
19. Salvador, S., Chan, P.: Toward accurate dynamic time warping in linear time and space. Intell. Data Anal. **11**(5), 561–580 (2007)
20. Schfer, P.: The boss is concerned with time series classification in the presence of noise. Data Min. Knowl. Discov. **29**(6), 1505–1530 (2015)
21. Shimodaira, H., Noma, K., Nakai, M., Sagayama, S.: Dynamic time-alignment kernel in support vector machine. In: Proceedings of the 14th International Conference on Neural Information Processing Systems: Natural and Synthetic, NIPS 2001, pp. 921–928. MIT Press, Cambridge (2001)
22. Wang, X., Mueen, A., Ding, H., Trajcevski, G., Scheuermann, P., Keogh, E.: Experimental comparison of representation methods and distance measures for time series data. Data Min. Knowl. Disc. **26**(2), 275–309 (2013)
23. Wei, L.Y.: A hybrid ANFIS model based on empirical mode decomposition for stock time series forecasting. Appl. Soft Comput. **42**, 368–376 (2016)
24. Xue, Y., Zhang, L., Tao, Z., Wang, B., Li, F.: An altered Kernel transformation for time series classification. In: Liu, D., Xie, S., Li, Y., Zhao, D., El-Alfy, E.-S.M. (eds.) ICONIP 2017 Part V. LNCS, vol. 10638, pp. 455–465. Springer, Cham (2017). https://doi.org/10.1007/978-3-319-70139-4_46
25. Zheng, Y., Liu, Q., Chen, E., Ge, Y., Zhao, J.L.: Time series classification using multi-channels deep convolutional neural networks. In: Li, F., Li, G., Hwang, S., Yao, B., Zhang, Z. (eds.) WAIM 2014. LNCS, vol. 8485, pp. 298–310. Springer, Cham (2014). https://doi.org/10.1007/978-3-319-08010-9_33

Planning and Optimization

Landmark-Biased Random Walk
for Deterministic Planning

Wei Wei[1,2], Chuang Li[1,3], Wei Liu[2], and Dantong Ouyang[1(✉)]

[1] College of Computer Science and Technology, Jilin University,
Changchun 130012, China
ouyd@jlu.edu.cn
[2] Center for Computer Fundamental Education, Jilin University,
Changchun 130012, China
[3] College of Computer, Jilin Normal University, Siping 136000, China

Abstract. Monte-Carlo random walk is a new wave of forward chaining search planning methods, which adopt more explorative search algorithms to handle heuristic deficiencies. During planning, random walks are used to explore the local neighborhood of a search state for action selection. In this paper, we propose a landmark-biased random walk planning method, where landmark is used to bias the random action selection by giving high probability to those actions that might achieve more landmarks. At each transition, bounded random walks, sequences of randomly selected actions are run to explore the neighborhood of the search state under the guidance of unachieved landmarks. Local exploration using landmark-biased random walks can make fast transitions in the search space, which is quite significant in regions where heuristic values do not help much. The experiments on challenging domains show that landmark is helpful for random walk to overcome the problem of executing too many worthless actions. The landmark-biased random walk improves the runtime effectively and usually obtains the first plan with better quality.

Keywords: Planning · Monte-Carlo random walk · Landmark
Action selection

1 Introduction

Heuristic search is currently one of the most effective techniques for deterministic planning. Most of the state of art heuristic planners, such as Fast Forward (FF) [1], Fast Downward [2] and LAMA [3], use variations of heuristic functions to guiding the search. In the last decade, much research has been conducted in designing powerful heuristic estimators to guide the search towards a goal state. While these methods contribute to very fast performance in many IPC benchmarks, they can lead to serious inefficiencies where the heuristic values are misleading [4].

Recently, a new wave of heuristic planners tries to use more explorative search algorithms to handle heuristic deficiencies [5, 6]. Monte Carlo techniques have been proved to be viable for large scale problems and challenging planning domains [7, 8]. Monte-Carlo Random Walk (MRW) is an algorithm for deterministic planning that

Z.-H. Zhou et al. (Eds.): ICAI 2018, CCIS 888, pp. 155–165, 2018.
https://doi.org/10.1007/978-981-13-2122-1_12

uses random exploration of the local neighborhood of a search state for selecting a promising action sequence. This planning method combines the exploration power of fast random walks with the strength of the available heuristic functions. During the planning procedure, the available heuristic function is still the main guiding engine of the algorithm, and local explorations using random walks and fast transitions in the search space using long jumps are significantly helpful in regions such as plateaus where heuristic values do not help much.

Landmarks are facts/actions that are necessary to achieve the goal for a given planning task, the use of which is usually a key mechanism to guide the search for a plan. Previous works have successfully exploited landmarks to identify subgoals [9–11] or devise informed heuristics [12–14] in both optimal and satisficing planning. Recently, landmarks have also been applied to temporal planning to describe temporal information of both what must happen and when this must happen [15], and numeric planning problems to capture metric conditions that are necessarily achieved in every plan [16].

Landmarks can be regarded as elicitation of problem's structure, which has an obvious impact on planning efficiency. It is still an important research trend to design more search strategies based on the exploitation of landmarks. In this paper, we suggest adapting landmarks in Monte Carlo random walk planning method to bias action selection. The main idea is to use the landmarks that are not achieved yet to guide the run of random walks. At each transition, sequences of randomly selected actions are built based on statistics from landmarks to bias the action selection towards promising actions that might achieve more landmarks.

The remainder of this paper is organized as follows: Sect. 2 briefly reviews the notation and background of planning task and landmark. Section 3 describes the architecture of Monte Carlo random walk planning and the probabilistic action selection in particular. Section 4 introduces the idea of landmark-biased random walk suggested in this paper and the detailed algorithms. Section 5 discusses the experiment results on the challenging domains from IPC competition. Section 6 gives concluding remarks and some potential future work.

2   Notation and Background

We consider planning task in the SAS+ formalism that can be automatically generated from its PDDL description.

Definition 1 (Planning task). A SAS+ planning task is given by a 5-tuple $\Pi = <V, A, s_0, G, C>$, where

- $V = \{v_1, \ldots, v_n\}$ is a set of state variables, each associated with a finite domain. Each complete assignment s to V is called a state;
- s_0 is the initial state;
- the goal condition G is a partial assignment to V;
- A is a finite set of actions, where each action a is a pair $<pre(a), eff(a)>$ of partial assignments to V called preconditions and effects, respectively.
- Each action $a \in A$ has a non-negative cost $C(a)$.

Action a is applicable in state s iff $pre(a) \subseteq s$. Apply action a in state s, denoted by $s(a)$, can change the value of each state variable v to $eff(a)[v]$ if $eff(a)[v]$ is specified. By $s(<a_1, ..., a_k>)$ we denote the state obtained from sequential application of $a_1, ..., a_k$ starting at state s. An action sequence $<a_1, ..., a_n>$ is a plan if $G \subseteq s(<a_1, ..., a_n>)$.

Definition 2 (Landmark). Given a planning task Π, a propositional logic formula ϕ is a landmark iff in each plan, ϕ is true at some time on the way to the goal.

Finding the complete set of landmarks for a planning task is as hard as solving it. Thus, most works approximate the landmark set using some kind of problem relaxation, with the hope that such a relaxation still contains useful structural information. In addition to knowing landmarks, it is also useful to know in which order they should be achieved. The way of discovering landmarks and their orderings is tangential to our contribution. In what follows, we assume that a planning task Π is given with a landmark graph $<L, O>$, where L is the landmark set for Π and O is a set of orderings over L.

3   Monte-Carlo Random Walk for Deterministic Planning

3.1   The Architecture of MRW Planning

Monte-Carlo Random Walk for deterministic planning uses a forward-chaining local search algorithm. The search engine uses fast random walks to explore the neighborhood of a search state and transitions through the search space by jumping to the states found in the local neighborhood. For each search state s_0, n bounded random walks, sequences of randomly select actions are run, which will obtain a relatively large set S of states in the neighborhood. Each state in S is an endpoint of a new random walk starting from s_0. All states in S, but no states along the walks are evaluated by a heuristic function since the heuristic evaluation is usually orders of magnitude slower than state generation. This enables the algorithm to get a large set of samples in a short period of time. After n random walks, the next state is switched to the endpoint with minimum heuristic value. The search episode terminates if either a goal state is found, or it fails to improve the minimum heuristic value for several jumps. After termination a new search episode starts with a restart mechanism.

MRW planning method carefully balances between exploitation and exploration to deal with both of them. When the heuristic evaluation is poor, it does not totally rely on the assumption about the heuristic function. Instead, it commits to an action sequence that can quickly escape from local minima area.

3.2   Probabilistic Action Selection

The main idea of MRW planning is to better explore the current search neighborhood. Inside the Arvand planner that adopts MRW planning, three variations of MRW methods are implemented. In pure random walks, all applicable actions of the current search state are equally likely to be explored. To address the problems of dead-end states and large average branching factor, two different pruning methods are experimented to avoid exploring all possible actions: Monte Carlo Helpful Actions (MHA) and Monte Carlo Deadlock Avoidance (MDA). Their common motivation is to use statistics from the earlier random walks to improve future random walk. This is done by biasing the action selection towards more promising actions and away from non-promising ones.

For current search state s, the probability $P(a, s)$ that action a is chosen among all applicable actions $A(s)$ is set to

$$P(a, s) = \frac{e^{Q(a)/\tau}}{\sum_{b \in A(s)} e^{Q(b)/\tau}} \tag{1}$$

The parameter τ is used to flatten the probability distribution. MHA and MDA compute $Q(a)$ differently. MHA defines $Q(a)$ as how often action a was a helpful action at an evaluated endpoint in all random walk so far. MDA defines $Q(a)$ through the number of times that action a has appeared in a failed random walks and tries to sample actions with higher failure rate less often.

4   Biasing the Random Walk with Landmarks

4.1   Action Selection Biased with Landmarks

Landmarks are inevitable elements that must be achieved on the refinement paths to any solution. Exploiting landmarks during planning is based on the intuition that humans might use. Given a planning problem, the relevant landmarks and orders can be represented by a directed graph called landmark graph in the pre-processing step. For each state s reached by a path π, two sets can be extracted based on the set L of all discovered landmarks: $Accepted(s,\pi)$ is the set of accepted landmarks and $ReqAgain(s,\pi)$ is the set of accepted landmarks that are required again. Landmark counting heuristic uses such information to approximate the goal distance as the estimated number of landmarks that are still needed to be achieved from s onwards. These landmarks are given by

$$L(s, \pi) := (L \backslash Accepted(s, \pi)) \cup ReqAgain(s, \pi). \tag{2}$$

The number $|L(s,\pi)|$ is then the heuristic value assigned to state s. The landmark counting heuristic is one choice of many heuristic candidates in MRW planning, which is used to evaluate the endpoints of random walks and guide long jumps in the search space.

Concerning the characteristic of landmarks, we propose to use landmarks in addition to the heuristic to improve performance, by biasing the action selection of random walks towards those actions that might achieve more landmarks. In most modern planners that use landmark information, the processing step usually builds some data structures for efficient landmark computation, which map propositions to the actions that achieve them or have them as preconditions. Such data structures are quite significant in this context since they provide a more promising way to determine probabilistic actions.

Definition 3 (Support action). Given a landmark ϕ and an action a, if there exists a propositional fact p such that $p \in eff(a)$ and $p \in \phi$, then a is called a support action for landmark ϕ.

While landmarks can be any formulas over facts, we restrict our attention to single propositional fact and disjunctions of facts. The above definition treats disjunctive landmarks as a set of single propositional facts. Following the definition of support action, we can achieve our pruning technique as a by-product of the landmark counting heuristic function at every endpoint of random walks. For an endpoint state s that need to be evaluated, $h = |L(s,\pi)|$ can be used to approximate the goal distance. Additionally, for all the landmarks appearing in $L(s,\pi)$, their support actions are collected. For this concept, we use $Q(a)$ to keep track of times that action a has been collected as a support action in the formula of the probability $P(a,s)$. The probabilities that each action is chosen among all the applicable is updated after every heuristic evaluation of the current random walk. We give priority to support actions of all the unachieved landmarks, which means such actions have high probability to be selected and redundant actions for solution could be avoided to some extent. This ingenious technique is meaningful for overcoming the problem that random walking usually executes many worthless actions and makes the solutions too long.

4.2   The Algorithm

For a given planning problem, a forward-chaining search adopting Monte Carlo random walk builds a chain of states $s_0 \rightarrow s_1 \rightarrow s_2 \rightarrow \dots \rightarrow s_n$ where s_0 is the initial state, s_n is a goal state and each transition $s_i \rightarrow s_{i+1}$ means jumping from s_i to s_{i+1} through an action sequence found by random walking in the local neighborhood of s_i.

Algorithm 1 Planning Procedure Using Monte-Carlo Random Walks

Input Initial State s_0, goal condition G, action set A and landmark
graph $<L, O>$
Output A solution plan

$s \leftarrow s_0$
$h_{min} \leftarrow h(s_0)$
$counter \leftarrow 0$
while s does not satisfy G **do**
    if counter>MAX_STEPS or DeadEnd(s) **then**
        $s \leftarrow s_0$ {restart}
        $counter \leftarrow 0$
    end if
    $s \leftarrow LandmarkBiasedRandomWalk(s, G, <L, O>)$
    if $h(s)<h_{min}$ **then**
        $h_{min} \leftarrow h(s_0)$
        $counter \leftarrow 0$
    else
        $counter \leftarrow counter+1$
    end if
end while
return the solution plan

Algorithm 1 shows an outline of the landmark-biased MRW planning procedure. In detail, for each search state s, a number of bounded random walks are run and the obtained states are evaluated. After that, the search engine jumps to the endpoint with minimum heuristic value and makes one transition. The search terminates successfully if a goal state is found. If the heuristic value does not improve within a given number of search steps, or it gets stuck in a dead–end state, the search fails. There is a restart mechanism in such cases.

Algorithm 2 Procedure of Landmark Biased Random Walk

Input current state s, goal Condition G and landmark graph $<L, O>$
Output s_{min} with minimum heuristic value after random walks

    $h_{min} \leftarrow INF$
    $s_{min} \leftarrow NULL$
    $L' = CollectLandmarks(s, G, <L, O>)$
    $A_{L'} = GenerateSupportActions(L')$
    $UpdateActionProbability(A_{L'})$
    for $i \leftarrow 1$ to NUM_WALK **do**
        $s' \leftarrow s$
        for $j \leftarrow 1$ to $LENGTH\_WALK$ **do**
            $A \leftarrow ApplicableActions(s')$
            if $A = \emptyset$ **then**
                break
            end if
            $a \leftarrow RandomSelectFromSupported(A)$
            $s' \leftarrow apply(s', a)$
            if s' satisfies G **then**
                return s'
            end if
        end for
        if $h(s') < h_{min}$ **then**
            $s_{min} \leftarrow s'$
            $h_{min} \leftarrow h(s')$
        end if
    end for
    if $s_{min} = NULL$ **then**
        return s
    else
        return s_{min}
    end if

Algorithm 2 illustrates how random walks are biased by landmarks to make long jumps in the search space. The parameters *LENGTH_WALK* and *NUM_WALK* are necessary. A walk stops at a goal state, a dead-end, or when its length reaches a bound *LENGTH_WALK*. The algorithm terminates either when a goal state is reached, or after *NUM_WALK* walks. The endpoint state with minimum h-value and the action sequence leading to it are returned. Before the random walk procedure, the set L' of unachieved landmarks is collected. Then the set $A_{L'}$ of support actions for all the landmarks in L' is generated and used to update probabilistic action values. After this process, the random walks from state s get started. At each step during walking, an action a is selected according to the probability biased by landmark information, where applicable actions are not equally likely to be explored, the actions that support some landmark are more promising to be selected.

5 Experiments

The landmark-biased MRW planning method is implemented on top of Arvand planner, which supports full PDDL description and uses the SAS+ formalism of planning problem. Three methods are tested: Pure random walk that explored all applicable actions equally, MHA random walk that bias action selection with help actions produced by FF heuristic, Landmark-biased random walk suggested in this paper, notated by LRW. In all experiments, the time and memory limits were 1200 s and 4 GB respectively for each task, running on a 2.30 GHz Intel Core CPU.

The parameters are shown in Table 1. All these values were determined experimentally or the same as Arvand planner. Each run uses an initial bound *LENGTH_WALK*, and successively extends it by iterative deepening. The experiments use a more flexible *acceptable progress* mechanism than using a fixed *NUM_WALK* to stop exploration once a state with small enough h-value is reached. The setting $\alpha = 0.9$ put more emphasis on recent *progress*. The parameters *EXTENDING_PERIOD* and *EXTENDING_RATE* control the length extending schedule. The search fails if h-value does not improve within a given number *MAX_STEPS*. The parameter τ stretches or flattens the probability distribution.

Table 1. Experimental parameters

Parameters	Values
LENGTH_WALK	10
NUM_WALK	1000
EXTENDING_PERIOD	300
EXTENDING_RATE	1.5
MAX_STEPS	7
α	0.9
τ	10

Experiments are tested on some challenging domains from previous IPCs[1]: Sanalyzer, Transport, Openstack, Visitall. Each domain contains a diverse set of problems. For each task, the average runtime and plan quality over ten runs are given if the task can be solved within ten runs. The results are summarized in the following figures.

Figure 1 illustrates how landmark contributes to the runtimes of random walks in these domains. Missing data dots indicate that the corresponding task is not solved within ten runs with 1200 s limit. The results show that landmark-biased random walk performs better than the other two methods. Pure random walk without any fall-back strategy is the weakest. MHA gives priority to actions that have been selected as a helpful action at an endpoint. But, helpful actions would not be computed at each step and thus MHA works only one time at the beginning of a random walk. Landmarks bias the action selections in

---

[1] IPC 2011, http://www.plg.inf.uc3m.es/ipc2011-deterministic/
IPC 2014, http://helios.hud.ac.uk/scommv/IPC-14/.

the whole process of random walking, which could prune much bigger irrelevant regions of the search and thus contribute a much more efficient runtime.

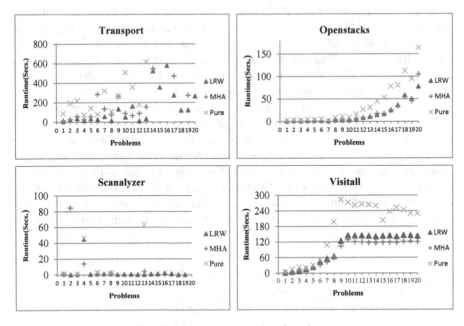

Fig. 1. Experimental results of runtimes

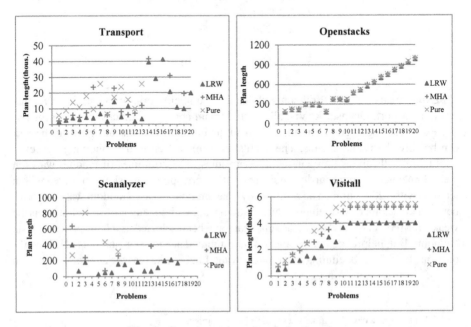

Fig. 2. Experimental results of plan lengths

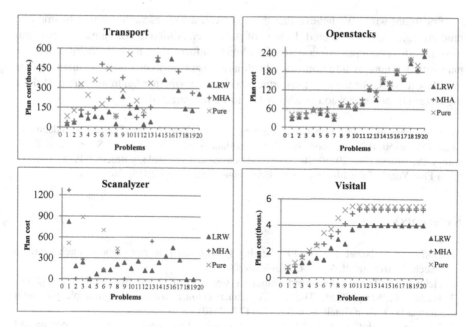

Fig. 3. Experimental results of plan costs

Regarding plan quality, almost all the current planners use a post-processing tool to remove useless loops in the plan to improve the plan quality. Also, anytime search engine continues to search for better solution after finding a first plan until it has exhausted the memory limit. Nevertheless, the quality of the initial plan is still quite important and determines the runtime directly since it is the cost bound of the subsequent restart procedures. Thus, we only execute the basic run to generate the first plan in the experiments. Both the length and the cost of plan are recorded to compare plan quality. The results illustrated in Figs. 2 and 3 show that the landmark-biased random walk has a huge improvement in both plan length and plan cost for most tested tasks, which means landmarks have an excellent guidance power to direct random walk closing goals through a much shorter and lower cost trajectory.

6 Conclusion

This research shows how landmark can improve the power of Monte Carlo random walk planning method. Support actions for unachieved landmarks are collected in our method and assigned higher probabilities to be selected. The search engine explores the local search space based on such landmark-biased action selection probabilities. For efficiency, support actions are only collected at endpoints as a byproduct when the planner uses the landmark heuristic evaluation. It is a successful use of landmark in addition to heuristic evaluation. The research in this paper indicates that landmark has a significant guidance power for random walk planning method since it helps to exploring the search space in a more reasonable way to handle heuristic deficiencies.

For future work, we believe that there are still many research avenues to improve random walk planning method. In term of parameters, different configurations perform well in different types of domain. Landmark information may be helpful for the parameter learning by intuition, since landmarks also contribute different enhancements in different tasks. For the smart restarting mechanism, whether landmarks are useful for the restarting point decision is also worth to be investigated.

Acknowledgements. The research is supported by National Nature Science Foundation of China (No. 61502197), Development Plan Foundations of Jilin Provincial Science and Technology Department (No. 20150520058JH) and Science and Technology Research Plan Project of "13th Five-Year" of the Jilin Provincial Education Department ([2016], No. 434).

References

1. Hoffmann, J., Nebel, B.: The FF planning system: fast plan generation through heuristic search. J. Artif. Intell. Res. **14**, 253–302 (2001)
2. Helmert, M.: The fast downward planning system. J. Artif. Intell. Res. **26**, 191–246 (2006)
3. Richter, S., Westphal, M.: The LAMA planner: guiding cost-based anytime planning with landmarks. J. Artif. Intell. Res. **39**, 127–177 (2010)
4. Nakhost, H., Hoffmann, J., Müller, M.: Improving local search for resource-constrained planning. Technical report TR 10-02, Department of Computing Science, University of Alberta, Edmonton, Alberta, Canada (2010)
5. Coles, A., Fox, M., Smith A.: A new local search algorithm for forward-chaining planning. In: Proceedings of the 17th International Conference on Automated Planning and Scheduling (ICAPS 2007), pp. 89–96 (2007)
6. Kartal, B.: Stochastic planning in large search spaces. In: Proceedings of 25th International Joint Conference on Artificial Intelligence (IJCAI 2016), pp. 4000–4001 (2016)
7. Nakhost, H., Müller, M.: Monte-Carlo exploration for deterministic planning. In: Proceedings of the 21st International Joint Conference on Artificial Intelligence (IJCAI 2009), pp. 1766–1771 (2009)
8. Cazenave, T., Shffidine, A., Schofield, M., Thielscher, M.: Nested Monte Carlo search for two-player games. In: Proceedings of the 30th AAAI Conference on Artificial Intelligence (AAAI 2016), pp. 687–693 (2016)
9. Hoffmann, J., Porteous, J., Sebastia, L.: Ordered landmarks in planning. J. Artif. Intell. Res. **22**, 215–278 (2004)
10. Vernhes, S., Infantes, G., Vidal, V.: Problem splitting using heuristic search in landmark orderings. In: Proceedings of the 23rd International Joint Conference on Artificial Intelligence (IJCAI 2013), pp. 2401–2407 (2013)
11. Bryce, D.: Landmark-based plan distance measures for diverse planning. In: Proceedings of the 24th International Conference on Automated Planning and Scheduling (ICAPS 2014), pp. 56–64 (2014)
12. Karpas, E., Domshlak, C.: Cost-optimal planning with landmarks. In: Proceedings of the 21st International Joint Conference on Artificial Intelligence (IJCAI 2009), pp. 1728–1733 (2009)
13. Domshlak, C., Katz, M., Lefler, S.: Landmark-enhanced abstraction heuristics. Artif. Intell. **189**, 48–68 (2012)

14. Bonet, B., Briel, M.: Flow-based heuristics for optimal planning: landmarks and merges. In: Proceedings of the 24th International Conference on Automated Planning and Scheduling (ICAPS 2014), pp. 47–55 (2014)
15. Karpas, E., Wang, D., Williams, B.C., et al.: Temporal landmarks: what must happen, and when. In: Proceedings of the International Conference on Automated Planning and Scheduling (ICAPS 2015), pp. 138–146 (2015)
16. Scala, E., Haslum, P., Magazzeni, D., et al.: Landmarks for numeric planning problems. In: Proceedings of 26th International Joint Conference on Artificial Intelligence (IJCAI 2017), pp. 4384–4390 (2017)

Gaussian Cauchy Differential Evolution for Global Optimization

Qingke Zhang[1,2](✉), Huaxiang Zhang[1,2](✉), Bo Yang[3], and Yupeng Hu[4]

[1] School of Information Science and Engineering, Shandong Normal University,
Jinan, China
`tsingke@hotmail.com, zhanghx@hotmail.com`
[2] Postdoctoral Research Station of Management Science and Engineering,
Shandong Normal University, Jinan, China
[3] Shandong Provincial Key Laboratory of Network based Intelligent Computing,
University of Jinan, Jinan, China
`yangbo@ujn.edu.cn`
[4] School of Computer Science and Technology, Shandong University, Jinan, China
`huyupeng@sdu.edu.cn`

Abstract. Differential evolution (DE) has been proven to be a powerful and efficient stochastic search technique for global numerical optimization. However, choosing the optimal control parameters of DE is a
time-consuming task because they are problem depended. DE may have
a strong ability in exploring the search space and locating the promising
area of global optimum but may be slow at exploitation. Thus, in this
paper, we propose a Gaussian Cauchy differential evolution (GCDE). It
is a hybrid of a modified bare-bones swarm optimizers and the differential evolution algorithm. It takes advantage of the good exploration
searching ability of DE and the good exploitation ability of bare-bones
optimization. Moreover, the parameters in GCDE are generated by the
function of Gaussian distribution and Cauchy distribution. In addition,
the parameters dynamically change according to the quality of the current search solution. The performance of proposed method is compared
with three differential evolution algorithms and three bare-bones technique based optimizers. Comprehensive experimental results show that
the proposed approach is better than, or at least comparable to, other
classic DE variants when considering the quality of search solutions on
a set of benchmark problems.

Keywords: Evolutionary computation · Differential evolution
Bare-bones particle swarm optimization · Adaptive parameters
Global optimization

Supported by the National Natural Science Foundation of China (61572298,
61772322, 61573166) and the Key Research and Development Foundation of Shandong Province (2017GGX1011).

© Springer Nature Singapore Pte Ltd. 2018
Z.-H. Zhou et al. (Eds.): ICAI 2018, CCIS 888, pp. 166–182, 2018.
https://doi.org/10.1007/978-981-13-2122-1_13

1 Introduction

Differential evolution (DE) proposed by [24], is a powerful and efficient optimizer. It is a population-based stochastic optimization algorithm for global optimization in continuous search space. Due to its simplicity, effectiveness and robustness, DE has been successfully applied in various optimization areas and many real-world applications, such as the field of artificial neural networks [1], pattern recognition [2], robotics and expert systems [6], image processing [5], and bioinformatics [20]. DE is a typical evolutionary algorithm (EA) that uses the mutation, crossover, and selection operators to evolve a population in solving a given optimization problems. The evolution process of DE starts from an initial population contained NP individuals in the search space and each individual represents a candidate solution of the problem. Then these solutions are selected as parents to generate offspring individuals (solutions) by mutation and crossover operation. The performance of DE depends on the control parameters (i.e., population size NP, scaling factor F and crossover probability C_r) and the trial vector generation strategy (i.e., mutation and crossover operators). After that, the new population is taken as the current population for next evolution operations. When the trial vector generation strategy is determined, the search performance of DE is sensitive to the choice of scale parameter and the probability of crossover. Generally, the optimal settings of these parameters are problem-dependent. Thus, it is often necessary to perform a time-consuming trial-and-error search procedure in order to search for the most appropriate strategy and to obtain the optimal results. A comprehensive survey of work concerning parameter control in evolutionary algorithms is analyzed in [14]. Another problem of DE is that it may gradually stop generating successful solutions when the population converged to a fixed local optimum point. This situation is commonly referred to as stagnation [17].

To solve these problems, researchers have proposed many empirical guidelines to improve the performance of DE. These methods can be classified into three groups. The first method is DE with adaption of strategy or control parameters. Parameter adaptation is adopted widely in DE variants, these adaption mechanisms contain the deterministic, adaptive and self-adaptive control. The typical DE improving algorithms are the self-adaptive DE (SaDE) [22], jDE [3], JADE [33], fuzzy adaptive DE (FADE) [16], success-history based parameter adaptation DE (SHADE) [25]. The second approach is DE with adaptive population size. Compared to the huge works on the adaption of F, C_r or offspring generation strategy, there are still few works have been investigated to the population size NP adaption. In [4] Brest and Maučec proposed a self-adaptive DE, jDE-scop, which employs three strategies and a population size reduction mechanism. Instead of monotonically reducing the size of NP, [28] proposed an population adaption according to the population diversity. The method with population reduction was also adopted in [13,32]. Apart from the former two approaches, the third schemes is hybridizing other techniques or algorithms with DE. To tackle issues of population premature convergence and stagnation, [29] have designed an auto-enhanced population diversity (APED) mechanism to enhance the population diversity. [12] proposed an effective and efficient successful-parent-selecting (SPS) framework to adapts the selection of parents by storing successful solutions into an archive during mutation and crossover operation. Recently, some of the novel strategies have been introduced by incorporating DE with other EAs, such as PSO with DE [19], BBPSO with DE [18,26], artificial bee colony algorithm(ABC) with DE [11], Gaussian bare-bones DE (GBDE) [26] etc. A comprehensive survey and advances of DE can be found in [8,9]. The above mentioned hybrid mechanism motivated us to investigate whether the DE performance can be improved by hybridizing the bare-bones search technique with the scaling factor F and crossover probability C_r changed adaptively.

The rest of this paper is organized as follows. In Sect. 2, the classical DE algorithm is briefly introduced. The detailed of our proposed approach is described in Sect. 3. Experimental results and analyses are presented in Sect. 4. Finally, conclusions and future work are given in last section.

2   De Algorithm

In this section, we briefly introduced the basic DE algorithm. Similar to other EAs, DE is a population-based search method that can be used to minimize an objective function by evolving a population with NP individuals on D-dimensional parameter vectors. As shown in (1), the population P_G contains NP individuals and each individual encode one of the candidate solutions:

$$P_G = \{\mathbf{x}_{i,G} | \mathbf{x}_{i,G} = (x_{i,G}^1, x_{i,G}^2, \cdots, x_{i,G}^D)^T\} \tag{1}$$

where $i = 1, 2, \cdots, NP$, G denotes the generation index, $\mathbf{x}_{i,G}$ represents the i-th individual in current population. The initial population should cover the entire search space by uniformly randomizing the initial individuals within the search space. For the value of the j-th dimension in individual i at the generation $G = 0$ can be initialized as follows in 2.

$$x_{i,0}^j = x_{min}^j + rand(0,1) \cdot (x_{max}^j - x_{min}^j) \tag{2}$$

where $rand(0,1)$ is a uniformly distributed random variable in the range $[0,1]$ and x_{max}^j, x_{min}^j are the maximum and minimum search bounds of j-th dimension, respectively. After initialization of the whole population, DE enters an evolutionary process by the operation of mutation, crossover, and selection.

2.1   Mutation

In each generation G, the mutation operation is applied to generate a mutant vector $\mathbf{v}_{i,G}$ (or called donor vector) with respect to each target vector $\mathbf{x}_{i,G}$ in current population. The performance of DE is very sensitive to the mutation operator. In order to classify the variants with different mutation operators, the notation "$DE/x/y/z$" is adopted, where x specifies the target vector to be mutated, y is the number of difference vectors used, and z represents the crossover scheme, such as exponential (exp), binomial (bin). Some well-known and most frequently used DE mutation strategies are listed as follows:

1. "*DE/rand/1*"

$$\mathbf{v}_{i,G} = \mathbf{x}_{r_1,G} + F \cdot (\mathbf{x}_{r_2,G} - \mathbf{x}_{r_3,G}). \tag{3}$$

2. "*DE/rand/2*"

$$\mathbf{v}_{i,G} = \mathbf{x}_{r_1,G} + F \cdot (\mathbf{x}_{r_2,G} - \mathbf{x}_{r_3,G})$$
$$+ F \cdot (\mathbf{x}_{r_4,G} - \mathbf{x}_{r_5,G}). \tag{4}$$

3. "*DE/best/1*"

$$\mathbf{v}_{i,G} = \mathbf{x}_{best,G} + F \cdot (\mathbf{x}_{r_1,G} - \mathbf{x}_{r_2,G}). \tag{5}$$

4. "*DE/best/2*"

$$\mathbf{v}_{i,G} = \mathbf{x}_{best,G} + F \cdot (\mathbf{x}_{r_1,G} - \mathbf{x}_{r_2,G})$$
$$+ F \cdot (\mathbf{x}_{r_3,G} - \mathbf{x}_{r_4,G}). \tag{6}$$

where the indices r_1, r_2, r_3 and r_4 denote the individual's index, which are randomly chosen from the set $\{1, 2, \cdots, NP\}$ and $r_1 \neq r_2 \neq r_3 \neq r_4 \neq i$, so that the size of NP must be at least four; $x_{best,G}$ is the best individual vector with the best fitness among the current population at generation G. The scale factor F is a real number ($F \in [0,2]$) which is used to control the amplification of the difference vector.

2.2   Crossover

In order to enrich the population diversity, DE employs a crossover operation to generator a trial vector $\mathbf{u}_i$ (novel solution) by recombining the current vector $\mathbf{x}_i$ and the mutant vector $\mathbf{v}_i$. In general, the DE family of algorithms commonly uses two crossover methods exponential and binomial [8]. The trial vector $\mathbf{u}_{i,G} = (u_{i,G}^1, u_{i,G}^1, \cdots, u_{i,G}^D)$ of individual i at the G-th generation by binomial crossover can be expressed as follows.

$$u_{i,G}^j = \begin{cases} v_{i,G}^j, & \mathrm{rand}_j(0,1) \leq C_r \vee j = j_{rand} \\ x_{i,G}^j, & \text{otherwise.} \end{cases} \tag{7}$$

where $i = 1, 2, \cdots, NP, C_r \in (0,1)$ is the crossover probability, $rand_j(0,1)$ is a uniform random number in the interval (0,1) for individual's each dimension, and $j_{rand} \in 1, 2, \cdots, NP$ is a random chosen indices within $[1, NP]$ which ensures that the trial vector $\mathbf{u}_{i,G}$ contained at least one parameter from the mutant vector $\mathbf{v}_{i,G}$.

2.3   Selection

In order to decide whether the trial vector can survive to the next population, a greedy selection criterion is performed via comparing the current target vector $\mathbf{x}_{i,G} = (x_{i,G}^1, x_{i,G}^2, \cdots, x_{i,G}^D)$ and the trial vector $\mathbf{u}_{i,G} = (u_{i,G}^1, u_{i,G}^2, \cdots, u_{i,G}^D)$ in terms of the objective function fitness value. The vector with better fitness function value will be survived to the next generation. The selection procedure is defined as follows:

$$\mathbf{x}_{i,G} = \begin{cases} \mathbf{u}_{i,G}, & \text{if } f(\mathbf{u}_{i,G}) \leq f(\mathbf{x}_{i,G}) \\ \\ \mathbf{x}_{i,G}, & \text{otherwise.} \end{cases} \tag{8}$$

where the equality "$\leq$" can help to navigate the flat portions of a fitness landscape and to reduce the possibility of population becoming stagnated [9].

3   GCDE Algorithm

In this section, the details of proposed algorithm GCDE are described. In GCDE, a novel mutant strategy is designed by combining a modified bare bones particle swarm optimizer and "DE/best/2" mutant strategy to generate the trial vector. The novel approach takes advantage of the good exploration searching ability of DE and the good exploitation ability of bare-bones optimization. Besides, the scaling parameter F and crossover probability Cr of GCDE are generated by Gaussian distribution function and Cauchy function, respectively. The parameters dynamically changed according to the quality of current search solution. The details of the proposed approach are described in the following part.

3.1   Bare-Bones Mutant Strategy

Particle swarm optimization (PSO) is a swarm intelligence-based stochastic search method inspired by simulate the social behavior of birds flocking and fish schooling. In PSO, each particle represents a candidate solution of the optimization problem and can be expressed by two vectors: position vector $\mathbf{x}_i = (x_i^1, x_i^2, \cdots, x_i^D)$ and velocity vector $\mathbf{v}_i = (v_i^1, v_i^2, \cdots, v_i^D)$, where D is the searching dimensional. During the searching evolution, each particle i maintains a memory of its historical personal best position $Pbest_i = (pbest_i^1, pbest_i^2, \cdots, pbest_i^D)$. These particles search along the entire space by moving with a corresponding velocity to find the global best position $Gbest = (gbest^1, gbest^2, \cdots, gbest^D)$. For particle i, the value of j-th dimension will be calculated by using the following formulas:

$$\begin{aligned} v_i^j &\leftarrow wv_i^j + c_1 r_1 (pbest_i^j - x_i^j) + c_2 r_2 (gbest^j - x_i^j), \\ x_i^j &\leftarrow x_i^j + v_i^j. \end{aligned} \tag{9}$$

where $i = 1, 2, \cdots, NP, j = (1, 2, \cdots, D), w$ is the inertial weight, c_1, c_2 are the acceleration coefficients and r_1, r_2 are uniformly distributed random real numbers in $[0, 1]$. Clerc and Kennedy [7] have theoretically proved that the particle converges to the weight average of its personal best and global best position when the evolving generation G trends to infinity. This can be formulated as:

$$\lim_{G \to +\infty} x_{i,G}^j = \frac{c_1 \cdot pbest_{i,G}^j + c_2 \cdot gbest_G^j}{c_1 + c_2} \tag{10}$$

This convergence limit formula in (10) provides the theoretically support for the Bare-bones particle swarm optimization algorithm (BPSO) [15] developed by Kennedy. In BPSO, the conventional velocity update formula is substitute by a Gaussian sampling distribution as follows:

$$x^j_{i,G+1} = N\left(\frac{gbest^j_G + pbest^j_{i,G}}{2}, |gbest^j_G - pbest^j_{i,G}|\right) \tag{11}$$

where the function N represents a Gaussian distribution with the value of mean $(gbest^j_G + pbest^j_{i,G})/2$ and standard deviation $|gbest^j_G - pbest^j_{i,G}|$; $x^j_{i,G+1}$ denotes the j-th dimension value of particle i at the next generation $G+1$. It is obvious to see that BPSO is a control parameter free and more compact. In [15], Kennedy also proposed a modified BPSO by using an alternative mechanism as follows:

$$x^j_{i,G+1} = \begin{cases} N(\mu^j_{i,G}, \sigma^j_{i,G}), & \text{if } rand_j(0,1) > 0.5 \\ \\ pbest^j_{i,G}, & \text{otherwise.} \end{cases} \tag{12}$$

where $rand_j(0,1)$ is a random value generated in the interval $[0,1]$; the mean value $\mu^j_{i,G} = (gbest^j_G + pbest^j_{i,G})/2$ and the standard deviation value $\sigma^j_{i,G} = |gbest^j_G - pbest^j_{i,G}|$. Inspired by the characteristic of control parameter free of BPSO, Wang $et\ al.$ [26] proposed the GBDE algorithm based on a variant of BPSO to generator the trial vector as follows:

$$x^j_{i,G+1} = N\left(\frac{gbest^j_G + x^j_{i,G}}{2}, |gbest^j_G - x^j_{i,G}|\right) \tag{13}$$

Inspired by the characteristic of the modified BPSO. This modified BPSO biases towards exploiting historical best positions. Omran $et\ al.$ [18] proposed another DE variants based on BPSO. The trial vector is generated in (14) as follows:

$$u^j_{i,G} = \begin{cases} p^j_{i,G} + r_{1,j}(x^j_{i_1,G} - x^j_{i_2,G}), \text{if } rand_j(0,1) > C_r \\ \\ pbest^j_{i_3,G}, \quad \text{otherwise} \end{cases} \tag{14}$$

where $r_{1,j}$ is a random real value in the interval $[0,1]$ for the j-th dimension, i_1, i_2 and i_3 are mutually three different indices ($i_1 \neq i_2 \neq i_3 \neq i$) which are randomly selected from the set $\{1, 2, \cdots, NP\}$; $p^j_{i,G}$ is formulated by:

$$p^j_{i,G} = r_{2,j} \cdot pbest^j_{i,G} + (1 - r^j_2) \cdot gbest^j_G \tag{15}$$

where r^j_2 is a uniformly distribute random value in the range $[0,1]$ for the j-th dimension of individual i. However, it still biased towards exploiting local best positions and may be easily fall into local optimal positions. Thus, we enhanced the global exploration ability by introducing a DE mutant strategy "DE/best/2/bin". We capitalized the strength of DE and then designed a novel mutant strategy. It generators the candidate trial vector $\mathbf{u}_{i,G} = (u_{i,G}^1, u_{i,G}^2, \cdots, u_{i,G}^D)$ as follows:

$$u_{i,G}^j = \begin{cases} p_{i,G}^j + rand_i \cdot (x_{r_{i,1},G}^j - x_{r_{i,2},G}^j), \\ \quad \text{if } rand_j(0,1) \leq Cr_i \\ \\ x_{best,G}^j + F \cdot (x_{r_{i,1},G}^j - x_{r_{i,2},G}^j) \\ \quad + F \cdot (x_{r_{i,3},G}^j - x_{r_{i,4},G}^j), \\ \text{otherwise.} \end{cases} \tag{16}$$

where $r_{i,1}, r_{i,2}, r_{i,3}$, and $r_{i,4}$ denotes individual i's parents indices and they are generated randomly in current population; $rand_i$ is a uniformly distribute random value in the range $[0,1]$; Each element of $\mathbf{u}_{i,G}$ can be generated by the formula (15) with a random selected $pbest_{r_i,G}$. The scaling parameter F in (16) is a real and constant factor $\in [0,2]$ that used to control the amplification of differential variation. In GCDE, it generated following a Gaussian distribution function. Details of the adaptive parameters in GCDE are described in the next subsection.

3.2 Adaptive Control Parameters

The performance of DE is very sensitive to the control parameters and they have a significant effect to the search quality. However, there is no fixed parameter settings that can achieve the best performance for all the optimization problems. Thus, various parameter control approach have been proposed for DE to dynamically adjust the involved parameters according to the search quality. Zaharie [31] have theoretically proved that DE could convergence to the global optimum in the long time limit if the scaling factor F can be transformed into a Gaussian random value. Thus, we adaptive the parameter $F_{i,G}$ of individual i at G-th generation from a Gaussian distribution. The crossover rate Cr is a probability $0 \leq Cr \leq 1$ and it is used to control the recombination of elements between trial vector and target vector. Generally, a large Cr in the range $(0.9, 1.0)$ can speed up convergence and be beneficial for multimodal or parameter depended non-separable problems [10]. In contrast, a small value in the range $(0, 0.2)$ of Cr is good for the unimodal or separable problems. The literature [21,23] suggests that a good choice of Cr is $(0.3, 0.9)$. Thus, we made a trade-off of these suggestions by randomly adapt the crossover probability $Cr_{i,G}$ of individual i at G-th generation from a Cauchy distribution. The parameters F and Cr of each individual are initialed randomly in the interval of $(0,1)$, then the parameter adaptively changed according to the current search quality by the formula (17) as follows:

$$F_{i,G+1} = \begin{cases} F_i, \text{if } f(\mathbf{u}_{i,G}) \leq f(\mathbf{x}_{i,G}) \\ Gaussian(0.5, 0.15), \quad \text{otherwise} \end{cases}$$

$$(17)$$

$$Cr_{i,G+1} = \begin{cases} Cr_i, \quad \text{if } f(\mathbf{u}_{i,G}) \leq f(\mathbf{x}_{i,G}) \\ Cauchy(0.5, 0.3), \qquad \text{otherwise} \end{cases}$$

where G represents the current generation; $\mathbf{u}_{i,G}$ and $\mathbf{x}_{i,G}$ denote the trial vector and current position of individual i at Gth generation, respectively. The parameters F_i and Cr_i of individual i will be regenerated when failed to update current target vector $\mathbf{x}_{i,G}$, otherwise, the parameters will be automatically inherited to the next generation. The pseudocode of GCDE algorithm is presented in Algorithm 1.

4   Experimental Results and Analysis

In this section, a comprehensive experiment is carried out. The experimental contents are divided into five parts. A detailed description of the test benchmark functions are presented in Sect. 4.1. An overall performance comparison between GCDE with the classical DE optimizers and the bare-bones optimizers are given in Sects. 4.2 and 4.3, respectively.

4.1   Test Benchmark Functions

In order to evaluate the performance of the proposed algorithm, twenty well-known test benchmark functions mentioned in [29,30]. The properties of these benchmark functions are listed in Table 1. These functions possess various characteristics such as unimodal, multimodal, shifted, rotated, separable and non-separable. In the numerical experiments, we have tested the 30-D and 50-D variables on these benchmark functions. In order to achieve a fair and reliable comparison, the population size is set to $NP = 50$ for 30-D problems, and $NP = 100$ for 50-D problems. The results reported in the experiments are the fitness averages and standard over 25 independent runs. Each run test was allowed $D \times 10^4$ fitness evaluations (FEs) of the objective function. The parameter settings for each compared algorithm are the same as its original paper. The performance of all the compared algorithms are evaluated in terms of mean and standard deviation of the function best mean solution error with $f(\mathbf{x}) - f(\mathbf{x}^*)$, where $\mathbf{x}^*$ is the global minimum of the objective function, and $\mathbf{x}$ is the best solution achieved by the test algorithm after a number function evaluations. When the solution error is smaller than $1E-6$, the evolution of current run is called a successful run. To get a statistical conclusion of the experimental results, the two-tailed Wilcoxon rank sum tests [27] with significance level $\alpha = 0.05$ is conducted to judge the significance of the performance between GCDE and its competitor. The cases are labeled with "+", "=", and "−" when the performance of GCDE is significantly better than, similar to and worse than the compared algorithm.

Algorithm 1. GCDE algorithm

Input: Population: M; Dimension: D; Genetation: T
Output: The best vector (solution) Δ

1: (initialization) ;
2: $t \leftarrow 0$;
3: **for** $i = 1$ to M **do**
4:    **for** $j = 1$ to D **do**
5:        $x_{i,t}^j = x_{min}^j + rand(0,1)\cdot(x_{max}^j - x_{min}^j)$;
6:    **end for**
7: **end for**
8: **while** $(|f(\Delta)| \geq \varepsilon)$ or $(t \leq T)$ **do**
9:    **for** $i = 1$ to M **do**
10:        **if** $(success_i == 0)$ **then**
11:            $F_{i,t} = Gaussian(0.5, 0.1)$;
12:            $Cr_{i,t} = Cauchy(0.5, 0.3)$;
13:        **end if**
14:        Select parents index r_1, r_2, r_3, r_4 of individual i;
15:        ▶ (*Mutation*)
16:        **for** $j = 1$ to D **do**
17:            $r = rand(0,1)$;
18:            $p = r * pbest_{r_i,t}^j + (1 - r) * gbest_t^j$;
19:            **if** $rand(0,1) \leq Cr_i$ **then**
20:                $v_{i,t}^j = p + rand(0,1) * (x_{r_1,t}^j - x_{r_2,t}^j)$
21:            **else**
22:                $v_{i,t}^j = gbest_t^j + F_i \cdot (x_{r_1,t}^j - x_{r_2,t}^j) + F_i \cdot (x_{r_3,t}^j - x_{r_4,t}^j)$;
23:            **end if**
24:        **end for**
25:        ▶ (*Crossover*)
26:        **for** $j = 1$ to D **do**
27:            $dim = rand()\%M$;
28:            **if** $rand(0,1) \leq Cr_i$ or $j == dim$ **then**
29:                $u_{i,t}^j = v_{i,t}^j$;
30:            **else**
31:                $u_{i,t}^j = x_{i,t}^j$;
32:            **end if**
33:        **end for**
34:        ▶ (*Selection*)
35:        **if** $f(\mathbf{u}_{i,t}) \leq f(\mathbf{x}_{i,t})$ **then**
36:            $success_i = 1$;
37:            $\mathbf{x}_{i,t} \leftarrow \mathbf{u}_{i,t}$;
38:            **if** $f(\mathbf{x}_{i,t}) < f(\Delta)$ **then**
39:                $\Delta \leftarrow \mathbf{x}_{i,t}$;
40:            **end if**
41:        **else**
42:            $success_i = 0$;
43:        **end if**
44:    **end for**
45:    $t \leftarrow t + 1$;
46: **end while**
47: **return** the best solution Δ;

Table 1. The properties and the formulas of the test benchmark functions.

Function Name	Test Benchmark Functions	Search Range	Global minimum
Sphere function	$F_1 = \sum_{i=1}^{D} x_i^2$	[-100,100]	0
Schwefel's function 1.2	$F_2 = \sum_{i=1}^{D} \sum_{j=1}^{i} x_j^2$	[-100,10]	0
Schwefel function 2.22	$F_3 = \sum_{i=1}^{D} |x_i| + \prod_{i=1}^{D} |x_i|$	[-10,10]	0
Rosenbrock function	$F_4 = \sum_{i=1}^{D} \left(100(x_i^2 - x_{i+1})^2 + (x_i - 1)^2\right)$	[-30,30]	0
Ackley's function	$F_5 = -20exp(-0.2\sqrt{\frac{1}{D}\sum_{i=1}^{D} x_i^2}) - exp(\frac{1}{D}\sum_{i=1}^{D} cos(2\pi x_i)))$ $+20 + e$	[-32,32]	0
Griewank's function	$F_6 = \sum_{i=1}^{D} \frac{x_i^2}{4000} - \prod_{i=1}^{D} cos(\frac{x_i}{\sqrt{i}}) + 1$	[-600,600]	0
Rastrigin's function	$F_7 = \sum_{i=1}^{D} (x_i^2 - 10cos(2\pi x_i) + 10)$	[-5.12,5.12]	0
Alpine Function	$F_8 = \sum_{i=1}^{D} |x_i sin(x_i) + 0.1 \times x_i|$	[-10,10]	0
Salomon Function	$F_9 = -cos(2\pi\sqrt{\sum_{i=1}^{D} x_i^2}) + 0.1 \times \sqrt{\sum_{i=1}^{D} x_i^2} + 1$	[-100,100]	0
Penalized function	$F_{10} = \frac{\pi}{D}\{10sin^2(\pi y_1) + A + (x_D - 1)^2\} + B$ $A = \sum_{i=1}^{D-1}(y_i - 1)^2[1 + 10sin^2(\pi y_{i+1})], B = \sum_{i=1}^{D} u(x_i, 5, 100, 4)$ $y_i = 1 + \frac{1}{4}(x_i + 1), u(x_i, a, k, m) = \begin{cases} k(x_i - a)^m & x_i > a, \\ 0, & -a \leq x_i \leq a \\ k(-x_i - a)^m. & x_i < -a. \end{cases}$	[-50,50]	0
shifted sphere function	$F_{11} = \sum_{i=1}^{D} z_i^2 + f_b, z = (x - o)$	[-100,100]	-450
Shifted Schwefel's problem 1.2	$F_{12} = \sum_{i=1}^{D}(\sum_{j=1}^{D} z_j^2) + f_b, z = (x - o)$	[-100,100]	-450
Shifted rotated high conditioned elliptic	$F_{13} = \sum_{i=1}^{D}(10^6)^{\frac{i-1}{D-1}} z_i^2 + f_b, z = M \cdot (x - o)$	[-100,100]	-450
Shifted Schwefel's problem 1.2 with noise in fitness	$F_{14} = (\sum_{i=1}^{D}(\sum_{j=1}^{i} z_j^2)) * (1 + 0.4 * |N(0,1)|) + f_b, z = (x - o).$	[-100,100]	-450
Schwefel?s problem 2.6 with global optimum on bounds	$F_{15} = max|A_i - B_i| + f_b,$	[-100,100]	-310
shifted Rosenbrock's function	$F_{16} = \sum_{i=1}^{D}(100(z_i^2 - z_i + 1)^2 + (z_i - 1)^2 + f_b, z = (x - o)$	[-100,100]	390
Shifted rotated Griewank's function without bounds	$F_{17} = \sum_{i=1}^{D}(\sum_{j=1}^{D} \frac{z_i^2}{4000} - \prod_{i=1}^{D} cos(\frac{z_i}{\sqrt{i}}) + 1 + f_b, z = (x - o)$	[0,600]	-180
Shifted rotated Ackley's function with global optimum on bounds	$F_{18} = -20exp(-0.2\sqrt{\frac{1}{D}\sum_{i=1}^{D} z_i^2}) - exp(\frac{1}{D}\sum_{i=1}^{D} cos(2\pi z_i)))$ $+20 + e + f_b, z = M \cdot (x - o)$	[-32,32]	-140
Shifted Rastrigin Function	$F_{19} = \sum_{i=1}^{D}(z_i^2 - 10cos(2\pi z_i) + 10) + f_b, z = (x - o)$	[-5,5]	-330
Schwefel's function 2.13	$F_{20} = \sum_{i=1}^{D}(|A_i - B_i(x)|)^2 + f_b,$ $A_i = \sum_{j=1}^{D}(a_{ij}sin\alpha_j + b_{ij}cos\alpha_j), B_i(x) = \sum_{j=1}^{D}(a_{ij}sinx_j + b_{ij}cosx_j)$	[-100,10]	-460

4.2 Experiment I: Comparison of GCDE with Classical de Optimizers

In this section, we compared the performance of GCDE with classical DE algorithms. such as DE/rand/1/, DE/rand/2/ and DE/best/2/. The algorithm DE/rand/1/ proposed by Price and Storn contains only one difference vector and has been mostly widely used. The algorithm DE/rand/2/ and DE/best/2/ contain two difference vectors and can generate more trial vectors to enrich the population diversity. Thus, they are apt for solving the multimodal problems. For common parameters F and Cr, these algorithms use the same settings, such as the population size, fitness evaluation times. As suggested by Price and Storn [24], the scaling factor F is set to 0.5 and the crossover probability Cr is set to 0.9. The numerical experiments are conducted on the 30-D and 50-D problems on the twenty benchmark functions from F_1 to F_{20}. The results achieved by DE/rand/1/bin, DE/rand/2/bin, DE/best/2/bin and GCDE are summarized in Tables 2 and 3 for 30-D and 50-D problems. The items "Mean", "St.D.", "SR" and "h" listed in the both tables represent the mean best fitness value, the corresponding standard deviation, success rate and the significance value, respectively.

The comparison results of 30-D and 50-D problems among GCDE and Other three classical DE algorithms, such as DE/rand/1, DE/rand/2, DE/best/2 are summarized as "$w/t/l$" at the last row in Tables 2 and 3. The symbol $w/t/l$ indicates that GCDE wins in the number of w benchmark functions, ties in the number of t functions and loses in the number of l functions. The best results among these algorithms are shown in **boldface**.

Table 2. Comparison between GCDE and DE algorithms on 30-D problems

Func.	DE/rand/1/bin				DE/rand/2/bin				DE/best/2/bin				GCDE		
	Mean	St. D.	SR%	h	Mean	St. D.	SR%	h	Mean	St. D.	SR%	h	Mean	St. D.	SR%
F_1	9.225E-44	3.951E-43	100%	+	8.085E-21	6.070E-21	100%	+	4.442E-92	1.805E-91	100%	+	**6.23E-212**	**0.00E+00**	100%
F_2	2.39E-45	1.17E-44	100%	+	1.06E-19	1.24E-19	100%	+	1.67E-90	5.72E-90	100%	+	**7.93E-210**	**0.00E+00**	100%
F_3	4.90E-30	2.24E-29	100%	+	1.32E-10	7.93E-11	100%	+	5.14E-48	9.30E-48	100%	+	**1.41E-115**	**1.16E-115**	100%
F_4	2.59E+01	1.72E+01	0%	+	1.14E+01	1.17E+01	0%	+	**3.84E-26**	**7.90E-26**	100%	-	2.58E-11	2.01E-11	100%
F_5	1.58E-11	7.72E-11	100%	+	2.38E-11	9.59E-12	100%	+	4.14E-15	6.96E-16	100%	-	6.84E-15	1.42E-15	100%
F_6	7.09E-03	1.03E-02	56%	+	**0.00E+00**	**0.00E+00**	100%	=	7.88E-03	6.42E-03	32%	+	**0.00E+00**	**0.00E+00**	100%
F_7	2.91E+00	1.09E+00	0%	+	1.14E+01	1.61E+00	0%	+	**1.89E-01**	**7.98E-01**	72%	-	1.99E+00	8.90E-01	20%
F_8	2.40E-01	4.90E-02	0%	+	**1.79E-01**	**3.78E-02**	0%	-	4.32E-01	9.26E-02	0%	+	2.00E-01	2.73E-12	0%
F_9	1.65E-32	4.05E-33	100%	+	4.82E-22	4.50E-22	100%	+	2.39E-32	1.72E-32	100%	+	**1.57E-32**	**0.00E+00**	100%
F_{10}	3.17E-02	2.55E-02	0%	+	4.82E-01	1.42E-01	0%	+	4.75E-02	6.56E-02	12%	+	**4.13E-16**	**3.77E-16**	100%
F_{11}	**0.00E+00**	**0.00E+00**	100%	-	**0.00E+00**	**0.00E+00**	100%	-	**0.00E+00**	**0.00E+00**	100%	-	3.41E-14	2.78E-14	100%
F_{12}	1.14E-03	1.18E-03	0%	+	1.71E-01	9.35E-02	0%	+	3.41E-14	2.78E-14	100%	-	7.30E-08	5.09E-08	100%
F_{13}	3.82E+05	1.49E+05	0%	+	1.05E+06	5.30E+05	0%	+	**2.93E+05**	**7.00E+05**	0%	-	1.74E+06	9.25E+05	0%
F_{14}	5.12E+00	9.56E+00	0%	+	3.18E+01	3.79E+01	0%	+	8.62E-01	2.14E+00	0%	+	**3.82E-02**	**3.23E-02**	0%
F_{15}	2.04E+03	6.21E+02	0%	+	2.02E+03	1.09E+03	0%	+	2.20E+03	1.29E+03	0%	+	**1.37E+03**	**8.03E+02**	0%
F_{16}	5.74E+01	3.94E+01	0%	+	1.87E+01	1.95E+01	0%	+	2.08E-07	1.02E-06	96%	+	**1.46E-10**	**1.49E-10**	100%
F_{17}	2.00E-02	1.60E-02	12%	+	**3.35E-03**	**4.56E-03**	64%	-	1.98E-02	1.58E-02	16%	+	1.77E-02	1.37E-02	40%
F_{18}	2.09E+01	7.40E-02	0%	+	2.10E+01	5.17E-02	0%	+	2.09E+01	4.88E-02	0%	+	**2.09E+01**	**4.33E-02**	0%
F_{19}	3.24E+00	1.55E+00	0%	+	1.52E+01	2.09E+00	0%	+	**2.57E+00**	**2.30E+00**	0%	-	4.58E+00	7.96E-01	0%
F_{20}	5.26E+04	4.31E+01	0%	+	2.18E+05	3.36E+04	0%	+	1.60E+05	3.11E+04	0%	+	**6.84E+03**	**8.10E+03**	0%
w/t/l	19/0/1				16/1/3				13/0/7				-/-/-		

According to the results on 30-D and 50-D problems, it can be observed that GCDE achieves better results than compared algorithms on a majority of test benchmark functions such as $F_1, F_2, F_3, F_4, F_6, F_8, F_9, F_{10}, F_{14}, F_{15}, F_{16}, F_{18}$, and F_{20}. The function F_1 to F_4, F_{11} to F_{15} are unimodal functions, GCDE achieves the promising solutions with the highest accuracy. Specially, among these unimodal functions, the Rosenbrock's function can be regarded as a multimodal function when the search dimension is beyond 3, because it has a narrow valley between the perceived local optima and the global optimum, it is easy to find the valley but very difficult to convergence to the global optimum for most optimizers. GCDE and the DE/best/2/bin can find the proximate optimum solution efficiently with a high success rate. The reminder functions F_5 to F_{10} and F_{16} to F_{20} are multimodal functions. It is not easy to reach the global optimum as there are many local optima in the searching space. Thus, most of the algorithms can not achieve the global optimum or with a low success rate. GCDE still performs better on those functions than other algorithms on these problems and has a relative higher success rate than other DE algorithms.

Table 3. Comparison between GCDE and DE algorithms on 50-D problems

Func.	DE/rand/1/bin				DE/rand/2/bin				DE/best/2/bin				GCDE		
	Mean	St. D.	SR%	h	Mean	St. D.	SR%	h	Mean	St. D.	SR%	h	Mean	St. D.	SR%
F_1	1.435E-25	8.315E-26	100%	+	6.779E-03	2.518E-03	0%	+	2.373E-39	3.950E-39	100%	+	**6.83E-113**	**8.96E-113**	100%
F_2	2.04E-24	1.15E-24	100%	+	1.26E-01	3.49E-02	0%	+	4.16E-38	5.37E-38	100%	+	**4.68E-111**	**1.07E-110**	100%
F_3	2.53E-14	1.17E-14	100%	+	8.79E-02	1.47E-02	0%	+	1.36E-20	6.60E-21	100%	+	**1.31E-59**	**2.07E-59**	100%
F_4	3.56E+01	1.34E+00	0%	+	4.86E+01	6.39E-01	0%	+	1.73E+00	3.19E+00	0%	+	**7.24E-01**	**1.09E-02**	0%
F_5	5.73E-14	1.53E-14	100%	+	2.13E-02	5.36E-03	0%	+	**9.68E-15**	**2.84E-15**	100%	-	1.04E-14	2.13E-15	100%
F_6	**0.00E+00**	**0.00E+00**	100%	=	8.20E-03	5.42E-03	0%	+	2.46E-03	5.05E-03	80%	+	**0.00E+00**	**0.00E+00**	100%
F_7	**5.24E-01**	**2.67E-01**	0%	-	5.78E+01	3.27E+00	0%	-	4.32E+01	4.32E+00	0%	-	8.70E+01	9.18E+00	0%
F_8	1.43E-27	1.17E-27	100%	+	3.48E-04	1.58E-04	0%	+	1.31E-32	4.79E-33	100%	+	**9.42E-33**	**0.00E+00**	100%
F_9	9.92E-03	7.40E-03	4%	+	3.90E+00	4.11E-01	0%	+	2.49E+00	5.89E-01	0%	+	**1.50E-16**	**1.86E-16**	100%
F_{10}	**2.00E-01**	**8.74E-10**	0%	-	1.01E+00	1.05E-01	0%	+	4.20E-01	6.00E-02	0%	+	2.10E-01	3.00E-02	0%
F_{11}	**0.00E+00**	**0.00E+00**	100%	-	1.05E-02	5.10E-03	0%	+	3.41E-14	2.78E-14	100%	+	**5.68E-14**	**0.00E+00**	100%
F_{12}	7.32E+01	3.19E+01	0%	+	2.23E+04	2.88E+03	0%	+	**3.84E-01**	**5.64E-01**	0%	+	4.10E+02	1.59E+01	0%
F_{13}	3.85E+06	1.21E+06	0%	+	8.32E+07	1.31E+07	0%	+	**1.63E+06**	**8.02E+05**	0%	+	1.65E+07	8.54E+06	0%
F_{14}	2.76E+03	1.18E+03	0%	+	4.34E+04	6.54E+03	0%	+	**2.13E+03**	**1.20E+03**	0%	+	7.61E+03	2.75E+03	0%
F_{15}	**2.99E+03**	**8.67E+02**	0%	+	1.32E+04	1.02E+03	0%	+	3.73E+03	1.07E+03	0%	+	3.17E+03	6.34E+02	0%
F_{16}	4.99E+01	2.56E+01	0%	+	8.72E+01	4.10E+01	0%	+	2.93E+00	4.79E+00	12%	+	**2.84E+01**	**1.44E+00**	0%
F_{17}	4.69E-03	5.88E-03	12%	+	1.10E+00	3.49E-02	0%	+	5.17E-03	6.91E-03	64%	+	**2.95E-03**	6.77E-03	80%
F_{18}	2.11E+01	3.91E-02	0%	=	2.11E+01	3.19E-02	0%	=	2.11E+01	2.17E-02	0%	=	2.11E+00	1.97E-02	0%
F_{19}	6.05E-01	3.49E-01	0%	-	6.48E+01	3.62E+00	0%	-	**4.77E+01**	**4.62E+00**	0%	-	7.52E+01	9.80E+00	0%
F_{20}	7.57E+05	7.11E+04	0%	+	1.11E+06	2.17E+05	0%	+	8.25E+05	1.10E+05	0%	+	**1.73E+04**	**1.14E+04**	0%
w/t/l	11/2/7				17/1/2				12/1/7				-/-/-		

4.3 Experiment II: Comparison of GCDE with Bare-Bones Optimizers

In this section, we compared the performance of GCDE with three bare-bones technique optimizers, namely the bare bones particle swarm optimization, namely, the BBPSO algorithm [15], bare-bones differential evolution (BBDE) [18], Gaussian bare-bones differential evolution (GBDE) [26]. The numerical experiments are conducted on 30-D and 50-D problems with function F_1 to F_{20}. The test results are tabulated in Tables 4 and 5. The items "Mean", "St.D.", "SR" and "h" listed in the tables represent the mean best fitness value, the corresponding standard deviation, success rate and the significance value. The comparison results are summarized as "$w/t/l$" in the last row. The symbol $w/t/l$ indicates that GCDE wins in the number of w benchmark functions, ties in the number of t functions and loses in the number of l functions. The best result among these algorithms are shown in **boldface**.

According to the results on 30-D and 50-D problems, it can be seen that GCDE outperforms BBPSO, BBDE and GBDE on 17 functions, 12 functions and 18 functions for 30-D problems and 16 functions for each 50-D problems. For the unimodal problems F_1 to F_4, and F_{11} to F_{14}, GCDE and BBPSO exhibited the best performance. Besides, the good performance of GCDE on the unimodal problems is consistent with the previous experimental results. For the multimodal functions, F_5 to F_{10} and F_{15} to F_{20}, GCDE performs relative better than other three optimizers on a majority of functions. GBDE generate the mutant

Table 4. Comparison between GCDE and the bare-bones algorithms on 30-D problems

Func.	BBPSO				BBDE				GBDE				GCDE		
	Mean	St. D.	SR%	h	Mean	St. D.	SR%	h	Mean	St. D.	SR%	h	Mean	St. D.	SR%
F_1	2.048E-117	1.001E-116	100%	+	2.045E-105	2.316E-105	100%	+	5.53E-99	1.58E-98	100%	+	**6.23E-216**	**0.00E+00**	100%
F_2	1.12E-123	4.38E-123	100%	+	1.82E-106	2.49E-106	100%	+	1.90E-104	2.82E-104	100%	+	**7.93E-215**	**0.00E+00**	100%
F_3	2.13E-84	5.56E-84	100%	+	5.08E-58	2.87E-58	100%	+	2.97E-68	4.58E-68	100%	+	**1.41E-115**	**1.16E-115**	100%
F_4	1.68E+01	1.93E+01	0%	+	2.49E+01	1.12E+01	0%	+	4.57E+01	3.13E+01	0%	+	**2.58E-11**	**2.01E-11**	100%
F_5	9.09E-02	3.14E-01	0%	+	7.55E-15	0.00E+00	100%	+	7.83E-15	1.39E-15	100%	+	**6.84E-15**	**1.42E-15**	100%
F_6	1.32E-02	1.40E-02	32%	+	**0.00E+00**	**0.00E+00**	100%	=	5.92E-04	2.01E-03	92%	+	**0.00E+00**	**0.00E+00**	100%
F_7	3.35E+01	9.84E+00	0%	+	**3.98E-02**	**1.95E-01**	96%	-	6.49E+00	2.98E+00	0%	+	1.99E+00	8.90E-01	20%
F_8	1.74E-01	2.98E-01	48%	+	**1.57E-32**	**1.57E-32**	100%	+	4.15E-03	2.03E-02	96%	+	**1.57E-32**	**0.00E+00**	100%
F_9	1.72E-14	1.18E-14	100%	+	**1.21E-52**	**4.69E-52**	100%	-	5.08E-15	6.00E-15	100%	+	4.13E-16	3.77E-16	100%
F_{10}	2.96E-01	5.99E-02	0%	+	2.92E-01	2.71E-02	0%	+	2.20E-01	4.00E-02	0%	+	**2.00E-01**	**2.73E-12**	0%
F_{11}	5.91E-14	1.11E-14	100%	+	5.68E-14	0.00E+00	100%	+	5.46E-14	1.11E-14	100%	+	**3.41E-14**	**2.78E-14**	100%
F_{12}	1.69E-04	2.18E-04	0%	+	5.61E+01	1.53E+01	0%	+	2.26E-01	1.67E-01	0%	+	**7.30E-08**	**5.09E-08**	100%
F_{13}	3.75E+06	4.47E+06	0%	+	1.35E+07	4.30E+06	0%	+	5.83E+06	4.41E+06	0%	+	**1.74E+06**	**9.25E+05**	0%
F_{14}	1.10E+02	8.88E+01	0%	+	4.75E+03	9.52E+02	0%	+	5.48E+02	4.93E+02	0%	+	**3.82E-02**	**3.23E-02**	0%
F_{15}	5.47E+03	1.24E+03	0%	+	3.74E+03	4.17E+02	0%	+	3.64E+03	6.40E+02	0%	+	**1.37E+03**	**8.03E+02**	0%
F_{16}	2.78E+01	3.71E+01	0%	+	4.72E+01	3.47E+01	0%	+	5.14E+01	4.52E+01	0%	+	**1.46E-10**	**1.49E-10**	100%
F_{17}	1.43E-02	1.15E-02	16%	-	**1.05E-02**	**4.83E-03**	0%	-	2.09E-02	1.51E-02	4%	+	1.77E-02	1.37E-02	40%
F_{18}	2.09E+01	5.16E-02	0%	=	2.09E+01	7.13E-02	0%	=	2.09E+01	5.02E-02	0%	=	**2.09E+01**	**4.33E-02**	0%
F_{19}	3.24E+01	9.35E+00	0%	-	**3.98E-02**	**1.95E-01**	0%	-	6.81E+00	2.93E+00	0%	-	4.58E+00	7.96E-01	0%
F_{20}	1.80E+04	2.24E+04	0%	+	1.66E+04	5.85E+03	0%	+	9.75E+03	1.13E+04	0%	+	**6.84E+03**	**8.10E+03**	0%
w/t/l	17/1/2				12/2/4				18/1/1				-/-/-		

Table 5. Comparison between GCDE and the bare-bones algorithms on 50-D problems

Func.	BBPSO				BBDE				GBDE				GCDE		
	Mean	St. D.	SR%	h	Mean	St. D.	SR%	h	Mean	St. D.	SR%	h	Mean	St. D.	SR%
F_1	2.125E-44	6.353E-44	100%	+	1.522E-59	1.222E-59	100%	+	2.57E-39	5.15E-39	100%	+	**6.83E-113**	**8.96E-113**	100%
F_2	1.35E-44	2.00E-44	100%	+	2.57E-58	8.74E-59	100%	+	4.07E-39	7.61E-39	100%	+	**4.68E-111**	**1.07E-110**	100%
F_3	9.09E-35	2.04E-34	100%	+	3.16E-33	4.32E-34	100%	+	5.33E-29	9.83E-29	100%	+	**1.31E-59**	**2.07E-59**	100%
F_4	5.64E+01	3.42E+01	0%	+	4.35E+01	4.32E+00	0%	+	5.44E+01	2.58E+01	0%	+	**7.24E-01**	**1.09E-02**	100%
F_5	3.24E-14	6.36E-15	100%	+	2.74E-14	3.26E-15	100%	+	2.32E-14	3.26E-15	100%	+	**1.04E-14**	**2.13E-15**	100%
F_6	7.14E-03	6.65E-03	40%	+	**0.00E+00**	**0.00E+00**	100%	+	7.40E-04	2.22E-03	90%	+	**0.00E+00**	**0.00E+00**	100%
F_7	7.87E+01	1.40E+01	0%	-	**0.00E+00**	**0.00E+00**	100%	-	1.88E+01	7.92E+00	0%	-	8.70E+01	9.18E+00	20%
F_8	2.49E-02	3.05E-02	48%	+	**9.42E-33**	**1.37E-48**	100%	+	6.22E-03	1.87E-02	90%	+	**9.42E-33**	**0.00E+00**	100%
F_9	3.77E-14	1.25E-14	100%	+	**2.69E-07**	**7.88E-07**	90%	+	7.36E-15	4.47E-15	100%	+	**1.50E-16**	**1.86E-16**	100%
F_{10}	4.30E-01	7.81E-02	0%	+	4.40E-01	4.90E-02	0%	+	3.70E-01	4.58E-02	0%	+	**2.10E-01**	**3.00E-02**	0%
F_{11}	1.14E-13	3.60E-14	100%	+	7.39E-14	2.60E-14	100%	+	7.39E-14	2.60E-14	100%	+	**5.68E-14**	**0.00E+00**	100%
F_{12}	**3.79E+02**	**2.44E+02**	0%	-	9.54E+03	1.11E+03	0%	+	3.13E+03	7.79E+02	0%	+	4.10E+02	1.59E+01	0%
F_{13}	3.41E+07	2.92E+07	0%	+	5.02E+07	1.10E+07	0%	+	**1.06E+07**	**2.84E+06**	0%	-	1.65E+07	8.54E+06	0%
F_{14}	1.53E+04	5.38E+03	0%	+	3.23E+04	4.39E+03	0%	+	2.28E+04	5.62E+03	0%	+	**7.61E+03**	**2.75E+03**	0%
F_{15}	1.08E+04	2.23E+03	0%	+	8.98E+03	6.03E+02	0%	+	8.16E+03	1.08E+03	0%	+	**3.17E+03**	**6.34E+02**	0%
F_{16}	7.59E+01	4.31E+01	0%	+	7.90E+01	2.75E+01	0%	+	7.43E+01	4.32E+01	0%	+	**2.84E+01**	**1.44E+00**	100%
F_{17}	6.18E-03	7.14E-03	12%	+	8.43E-03	2.50E-02	0%	+	5.69E-03	7.56E-03	16%	+	**2.95E-03**	**6.77E-03**	80%
F_{18}	2.11E+01	4.28E-02	0%	=	2.11E+01	2.82E-02	0%	=	2.11E+01	5.67E-02	0%	=	**2.11E+01**	**1.97E-02**	0%
F_{19}	8.59E+01	1.86E+01	0%	+	**1.08E-13**	**1.71E-14**	100%	-	2.30E+01	5.72E+00	0%	-	7.52E+01	9.80E+00	0%
F_{20}	1.50E+05	1.06E+05	0%	+	1.04E+05	1.32E+04	0%	+	3.74E+04	3.50E+04	0%	+	**1.73E+04**	**1.14E+04**	0%
w/t/l	16/1/3				16/2/2				16/1/3				-/-/-		

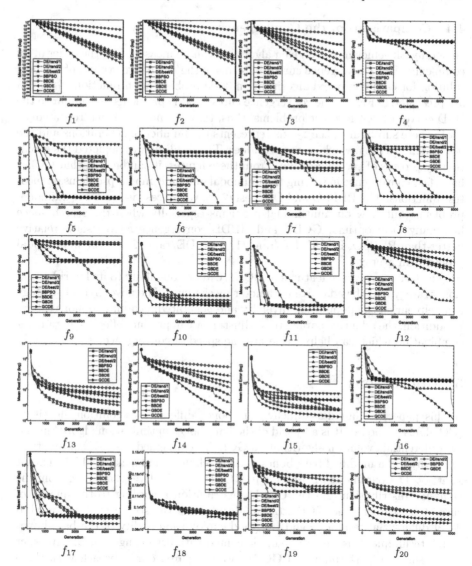

Fig. 1. The convergence curves of compared algorithms (DE/rand/1, DE/rand/2, DE/best/2, BBPSO, BBDE, GBDE, GCDE) with 30 variables on test functions F_1 to F_{20} (the x-axis represents the number of generations and the y-axis denotes the mean best fitness values)

vector by a modified mutant strategy and it performs better than the classical DE algorithm on multimodal function. Among the rotated and shifted problems, such as $F_{13}, F_{14}, F_{17}, F_{18}$, the results achieved by all compared algorithms are not comparable to the basic multimodal function. However, GCDE can still showed better performance with a relative higher success rate.

4.4   Comparisons of the Convergence

Generally, the efficiency of a given algorithm can be measured by the order of its complexity. The convergence curves in terms of the mean error values of the best solution for each compared algorithm on 30-D problem with function F_1 to F_{20} are presented in Fig. 1. The convergence cures of 50-D problems are similar to the 30-D curves except few test problems, thus, they are not be shown repeatedly.

In these sub-figures, the x-axis represents the number of generations and the y-axis denotes the mean best fitness values. These sub-figures depicted the evolutionary process of each algorithm. From these figures, it can be seen that GCDE has a fast convergence in solving the unimodal and multimodal problems on most test functions. The BBPSO and DE/best/2 convergences fast on basic unimodal problems, but slow on multimodal problems especially when the search space has many local optima. GCDE and GBDE convergence curves show relative better than BBPSO, BBDE, DE/rand/1/, and DE/rand/2/ on the multimodal problems in average.

The fast convergence features of GCDE can be contribute to its hybrid mutation strategy and the adaptive parameter mechanism. The novel hybrid strategy can help enrich the population diversity and generate much more donor vectors. In addition, the adaptive parameters adjustment can produce effective parameter combination which can help accelerate the speed of convergence.

5   Conclusions

In this paper, a Gaussian-Cauchy bare-bones differential evolution algorithm (GCDE) is proposed. It is a hybrid of the bare-bones optimization algorithm and the differential evolution algorithm. In GCDE, we designed a novel mutant strategy by hybrid a modified bare-bones search technique and a classical DE mutant strategy. The parameters F and Cr in GCDE are adaptively sampling from the Gaussian distribution and the Cauchy distribution according to the quality of current search solution. GCDE takes advantage of the strong exploration ability of differential technique and the good exploitation ability of the bare bones search technique. It is easy to be implemented with few changes on DE. In order to measure the performance of GCDE, we compared our approach with three classical DE algorithms and three bare-bones optimizer. Comprehensive experiments were conducted on twenty benchmark functions. The experimental results demonstrate that the proposed algorithm is an efficient and effective method in solving optimization problems and it can be comparable to other classic DE variants when considering the quality of the solutions on a set of benchmark problems. In future work, we will investigate the ensemble coevolving approach in GCDE and introduce other evolutionary search technique to enhance the global search performance.

References

1. Arce, F., Zamora, E., Sossa, H., Barrón, R.: Differential evolution training algorithm for dendrite morphological neural networks. Appl. Soft Comput. **68**, 303–313 (2018)
2. Bhadra, T., Bandyopadhyay, S.: Unsupervised feature selection using an improved version of differential evolution. Expert Syst. Appl. **42**(8), 4042–4053 (2015)
3. Brest, J., Greiner, S., Boskovic, B., Mernik, M., Zumer, V.: Self-adapting control parameters in differential evolution: a comparative study on numerical benchmark problems. IEEE Trans. Evol. Comput. **10**(6), 646–657 (2006)
4. Brest, J., Maučec, M.S.: Self-adaptive differential evolution algorithm using population size reduction and three strategies. Soft. Comput. **15**(11), 2157–2174 (2011)
5. Cai, Z.Q., Lv, L., Huang, H., Hu, H., Liang, Y.H.: Improving sampling-based image matting with cooperative coevolution differential evolution algorithm. Soft. Comput. **21**(15), 4417–4430 (2017)
6. Chen, C.H., Yang, S.Y.: Neural fuzzy inference systems with knowledge-based cultural differential evolution for nonlinear system control. Inf. Sci. **270**(2), 154–171 (2014)
7. Clerc, M., Kennedy, J.: The particle swarm-explosion, stability, and convergence in a multidimensional complex space. IEEE Trans. Evol. Comput. **6**(1), 58–73 (2002)
8. Das, S., Suganthan, P.N.: Differential evolution: a survey of the state-of-the-art. IEEE Trans. Evol. Comput. **15**(1), 4–31 (2011)
9. Das, S., Mullick, S.S., Suganthan, P.N.: Recent advances in differential evolution an updated survey. Swarm Evol. Comput. **27**, 1–30 (2016)
10. Gämperle, R., Müller, S.D., Koumoutsakos, P.: A parameter study for differential evolution. In: Advances in Intelligent Systems, Fuzzy Systems, Evolutionary Computation, vol. 10, pp. 293–298 (2002)
11. Gao, W., Liu, S.: Improved artificial bee colony algorithm for global optimization. Inf. Process. Lett. **111**(17), 871–882 (2011)
12. Guo, S.M., Yang, C.C., Hsu, P.H., Tsai, J.S.H.: Improving differential evolution with a successful-parent-selecting framework. IEEE Trans. Evol. Comput. **19**(5), 717–730 (2015)
13. Iacca, G., Mallipeddi, R., Mininno, E., Neri, F., Suganthan, P.N.: Super-fit and population size reduction in compact differential evolution. In: 2011 IEEE Workshop on Memetic Computing (MC), pp. 1–8. IEEE (2011)
14. Karafotias, G., Hoogendoorn, M., Eiben, A.E.: Parameter control in evolutionary algorithms: trends and challenges. IEEE Trans. Evol. Comput. **19**(2), 167–187 (2015)
15. Kennedy, J.: Bare bones particle swarms. In: Proceedings of the 2003 IEEE Swarm Intelligence Symposium, SIS 2003, pp. 80–87. IEEE (2003)
16. Liu, J., Lampinen, J.: A fuzzy adaptive differential evolution algorithm. Soft. Comput. **9**(6), 448–462 (2005)
17. Mohamed, A.W., Suganthan, P.N.: Real-parameter unconstrained optimization based on enhanced fitness-adaptive differential evolution algorithm with novel mutation. Soft. Comput. **22**(10), 3215–3235 (2018)
18. Omran, M.G.H., Engelbrecht, A.P., Salman, A.: Bare bones differential evolution. Eur. J. Oper. Res. **196**(1), 128–139 (2009)
19. Omran, M.G., Engelbrecht, A.P., Salman, A.: Differential evolution based particle swarm optimization. In: 2007 IEEE Swarm Intelligence Symposium, pp. 112–119. IEEE (2007)

20. Pie, M.R., Meyer, A.L.S.: The evolution of range sizes in mammals and squamates: heritability and differential evolutionary rates for low- and high-latitude limits. Evol. Biol. **44**(3), 347–355 (2017)
21. Price, K., Storn, R.M., Lampinen, J.A.: Differential Evolution: A Practical Approach to Global Optimization. Springer, Heidelberg (2006). https://doi.org/10.1007/3-540-31306-0
22. Qin, A.K., Huang, V.L., Suganthan, P.N.: Differential evolution algorithm with strategy adaptation for global numerical optimization. IEEE Trans. Evol. Comput. **13**(2), 398–417 (2009)
23. Ronkkonen, J., Kukkonen, S., Price, K.V.: Real-parameter optimization with differential evolution. In: Proceedings of IEEE CEC, vol. 1, pp. 506–513 (2005)
24. Storn, R., Price, K.: Differential evolution-a simple and efficient heuristic for global optimization over continuous spaces. J. Global Optim. **11**(4), 341–359 (1997)
25. Tanabe, R., Fukunaga, A.: Success-history based parameter adaptation for differential evolution. In: 2013 IEEE Congress on Evolutionary Computation, pp. 71–78. IEEE (2013)
26. Wang, H., Rahnamayan, S., Sun, H., Omran, M.G.: Gaussian bare-bones differential evolution. IEEE Trans. Cybern. **43**(2), 634–647 (2013)
27. Wilcoxon, F.: Individual comparisons by ranking methods. Biom. Bull. **1**(6), 80–83 (1945)
28. Yang, M., Cai, Z., Li, C., Guan, J.: An improved adaptive differential evolution algorithm with population adaptation. In: Proceedings of the 15th Annual Conference on Genetic and Evolutionary Computation, pp. 145–152. ACM (2013)
29. Yang, M., Li, C., Cai, Z., Guan, J.: Differential evolution with auto-enhanced population diversity. IEEE Trans. Cybern. **45**(2), 302–315 (2015)
30. Yu, W.J., et al.: Differential evolution with two-level parameter adaptation. IEEE Trans. Cybern. **44**(7), 1080–1099 (2014)
31. Zaharie, D.: Critical values for the control parameters of differential evolution algorithms. In: Proceedings of MENDEL, vol. 2002 (2002)
32. Zamuda, A., Brest, J.: Population reduction differential evolution with multiple mutation strategies in real world industry challenges. In: Rutkowski, L., Korytkowski, M., Scherer, R., Tadeusiewicz, R., Zadeh, L.A., Zurada, J.M. (eds.) EC/SIDE -2012. LNCS, vol. 7269, pp. 154–161. Springer, Heidelberg (2012). https://doi.org/10.1007/978-3-642-29353-5_18
33. Zhang, J., Sanderson, A.C.: JADE: adaptive differential evolution with optional external archive. IEEE Trans. Evol. Comput. **13**(5), 945–958 (2009)

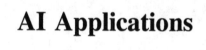

AI Applications

Improved Nearest Neighbor Distance Ratio for Matching Local Image Descriptors

Han Yan, Guohua Lv$^{(\boxtimes)}$, Xiaoqiang Ren$^{(\boxtimes)}$, and Xiangjun Dong

College of Information, Qilu University of Technology
(Shandong Academy of Sciences), No. 3501, Daxue Road, Changqing District,
Jinan 250353, China
{guohualv,renxq}@qlu.edu.cn

Abstract. This paper presents a novel matching strategy, called Improved Nearest Neighbor Distance Ratio, for matching local image descriptors. We first empirically analyze to what extent correspondences underlie the second nearest neighbor or even the third and so on. Based on the solid analysis, we propose to improve the widely-used Nearest Neighbor Distance Ratio (NNDR) by matching local descriptors not only based on the first nearest neighbor, but also making use of the second nearest neighbor appropriately. The proposed INNDR is evaluated against NNDR on a set of benchmark datasets. Our experiments will demonstrate that INNDR generally outperforms the traditional NNDR in both matching accuracy and recall vs 1-precision.

Keywords: NNDR · Improved NNDR · Local image descriptor
Keypoint matching

1 Introduction

Local image descriptors are essential in various applications of computer vision and image processing such as object recognition [1], image retrieval [2] and image registration [4,17,18]. In these applications, matching local descriptors is a crucial step which aims to find correct correspondences between two or multiple images.

In matching local image descriptors, the Nearest Neighbor Distance Ratio (NNDR) matching strategy has been extensively used and has shown its great effectiveness in a variety of applications [1,4,14,15,20–23]. The main idea of NNDR is searching for the nearest neighbor and meanwhile imposing a distance ratio between the first and the second nearest neighbor. This distance ratio facilitates the discrimination of the first nearest neighbor, thereby performing well in matching local image descriptors. In this paper we would like to ask

---

This paper is supported by Natural Science Foundation of Shandong Province, China (No. ZR2017LF005 and No. ZR2016FM14).

Z.-H. Zhou et al. (Eds.): ICAI 2018, CCIS 888, pp. 185–197, 2018.
https://doi.org/10.1007/978-981-13-2122-1_14

the following questions: (1) To what extent correct correspondences underlie the second nearest neighbor or even the third and so on? (2) If these correct correspondences are noteworthy, how to dig them out to better match local image descriptors?

This paper will address the aforementioned concerns and present a novel matching strategy, called Improved Nearest Neighbor Distance Ratio (INNDR), for matching local image descriptors. We first analyze the possibility of utilizing the second nearest neighbor or even more in the matching process. Based on the analysis, we then propose to improve NNDR by taking into account both the nearest neighbor and the second nearest neighbor. Our experiments show that the proposed INNDR generally outperforms the traditional NNDR in matching local image descriptors. Main contributions of this paper are twofold as follows.

i. Empirically analyzing the impact of the second nearest neighbor in matching local image descriptors;
ii. Proposing a novel matching strategy by improving the widely-used NNDR.

The rest of this paper is structured as follows. In Sect. 2, related work is briefly reviewed. Section 3 formulates the problem that underlies the nearest neighbor distance ratio matching strategy. In Sect. 4, the proposed matching strategy is presented, followed by a performance study in Sect. 5. The paper is concluded in Sect. 6.

2   Related Work

This section briefly reviews related work including local image descriptors, the NNDR matching strategy and Random Sample Consensus (RANSAC).

2.1   Local Image Descriptors

A local descriptor is designed to represent image information within a local region [1,3,4]. There exist a large number of local image descriptors in the literature, such as Scale Invariant Feature Transform (SIFT) [1], Partial Intensity Invariant Feature Descriptor (PIIFD) [5], Binary Robust Independent Elementary Features (BRIEF) [6] and Quaternionic Local Ranking Binary Pattern (QLRBP) [7], just to name a few. Among various kind of local descriptors, SIFT is the most popular one and it has been improved in many possible ways [3,8–12,18].

A SIFT-like descriptor is built by concatenating orientation histograms, where each histogram has a number of orientation bins. In traditional SIFT-like descriptors [1,9,11], gradient magnitudes (GM) of image pixels are summed up at each bin of an orientation histogram. In [13], gradient occurrences (GO) substitute GM in building local descriptors. Instead of summing GM, the number of locations where image gradients occur is counted when building local descriptors using GO. In contrast, GO are less susceptible to gradient changes between the corresponding parts of two images. These two types of gradient information, i.e., GM and GO, are analyzed and evaluated in [18]. The experiments on a set of benchmark datasets have shown that GO generally outperforms GM when registering images using SIFT-like descriptors.

2.2   Nearest Neighbor Distance Ratio

The NNDR matching strategy first appeared in [1] for the application of object recognition. In [1], object recognition is performed by first matching each keypoint of a query image to a database of keypoints extracted from training images. A large number of keypoints from the query image do not have any correct match in the training database because these keypoints arise from background clutter or are not detected in the training images. NNDR was proposed to discard these features in the process of matching keypoints.

To define NNDR, a few denotations are first given as follows.

i. D_q^r: a reference descriptor in query image;

ii. D_t^{1st}: the descriptor in target image which is the nearest to D_r (nearest neighbor);

iii. D_t^{2nd}: the descriptor in target image which is the second nearest to D_r (second nearest neighbor).

If the distance ratio between D_t^{1st} to D_q^r and D_t^{2nd} to D_q^r is below a threshold t, D_q^r and D_t^{1st} are matched, i.e.,

$$\frac{||D_q^r - D_t^{1st}||}{||D_q^r - D_t^{2nd}||} \leq t, \tag{1}$$

where $||D_q^r - D_t^{1st}||$ and $||D_q^r - D_t^{2nd}||$ are Euclidean distances of D_t^{1st} to D_q^r and D_t^{2nd} to D_q^r respectively [4,17], t is an empirical value, which equals 0.8. Due to the great effectiveness in feature matching, NNDR has been widely used since it was proposed [4,14–18,20–23].

2.3   Random Sample Consensus

Random sample consensus (RANSAC) [25] was proposed to find efficient data by estimating parameters of a mathematical model from a random dataset which contains outliers. For a given dataset, the main steps of RANSAC are as follows. First, the hypothetical inliers are chosen randomly from the input dataset. Second, a model is built to fit the hypothetical inliers. Third, the estimated model is utilized to test the remaining data. If the data fits this model in a acceptable error range, it is considered as consensus set. Last, once there is enough data in the consensus set, the estimated model is considered as the final solution. Otherwise, the hypothetical inliers are updated by iteratively adding new members.

3   Problem Formulation

In matching local image descriptors, it is the expectation of NNDR that the nearest neighbor of a query keypoint is its best match. However, it is not always the case. We have found that a keypoint in query image may spatially correspond to its second nearest neighbor, or even its third nearest neighbor afterwards. A real example is given in Fig. 1. The left image of Fig. 1(a) is a query image, where a reference descriptor is built within the region marked by a blue square. The descriptor is denoted as K_q^r. In the right image of Fig. 1(a), the two regions marked by a red square and a green square are the nearest neighbor and the second nearest neighbor to K_q^r. The keypoints centered at the red square and the green square are denoted as K_t^{1st} and K_t^{2nd} respectively. By visually observing Fig. 1(b) to (d), it is clear that $K_q^r \mapsto K_t^{2nd}$ is a correct match and $K_q^r \mapsto K_t^{1st}$ is an incorrect match.

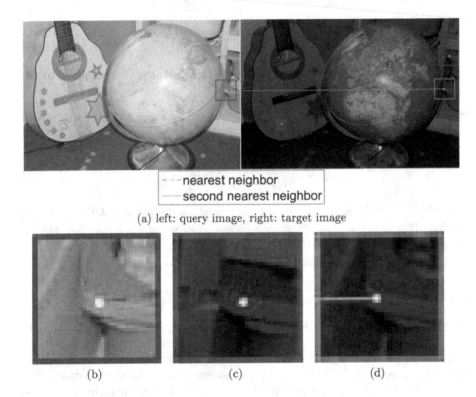

(a) left: query image, right: target image

Fig. 1. An example illustrating that the second nearest neighbor is a correct match and the nearest neighbor is not. (b): the local region marked in query image; (c): the nearest neighbor to (b) in target image; (d): the second nearest neighbor to (b) in target image. (Color figure online)

We have made a statistic on the number of correct matches underlying top 5 nearest neighbors, as shown in Fig. 2. This result was obtained by registering 253 image pairs which undergo various transformations. When generating the statistics in Fig. 2, GO-SIFT and GO-IS-SIFT (GO: Gradient Occurrences, IS: Improved Symmetric) [18] were used for feature description in mono-modal and multi-modal cases respectively. As shown in Fig. 2, the nearest neighbor is the main inhabitation of the ground-truth target keypoints. Interestingly, the second nearest neighbor plays an non-negligible role in achieving better matching results. In the statistic shown in Fig. 2, the number of correct matches underlying the second nearest neighbor is 14.48% of the number of correct matches underlying the nearest neighbor. To the best of our knowledge, there does not exist keypoint descriptor which is thoroughly invariant to complex, non-linear intensity variations between multi-modal images [19, 24]. This would give rise to uncertainties in the process of keypoint matching. Thus, it is possible that the nearest neighbor is not the best match of the query keypoint. Based on this analysis, we are inspired to explore the nearest $x(x \geq 2)$ neighbors in matching local image descriptors.

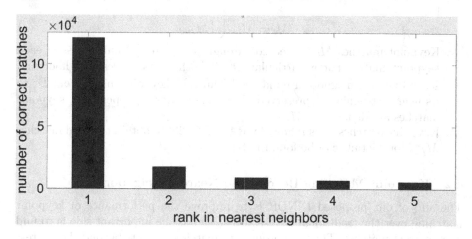

Fig. 2. A statistic on the number of correct matches underlying top 5 nearest neighbors.

4   Improved Nearest Neighbor Distance Ratio

This section elaborates our proposed INNDR matching strategy. Its overview is first given, followed by two key issues, i.e., keypoint matching beyond the nearest neighbor and refining keypoint matches using Random Sample Consensus (RANSAC) [25].

4.1   Overview of INNDR

First, local descriptors built in the query and target images are denoted as

$$\mathcal{D}_q = \{D_q^1, D_q^2, \ldots D_q^{N_q}\} \tag{2}$$

and

$$\mathcal{D}_t = \{D_t^1, D_t^2, \ldots D_t^{N_t}\}, \tag{3}$$

where N_q and N_t denote the number of descriptors in the query and target images respectively.

With $\mathcal{D}_q$ and $\mathcal{D}_t$ as the input, major steps of INNDR are performed as follows.

i. For each descriptor in query image, $D_q^i (1 \leq i \leq N_q)$, its Euclidean distance to any descriptor in target image, $D_t^j (1 \leq j \leq N_t)$, is calculated. All descriptors in target image are ranked by their Euclidean distances to D_q^i.

ii. D_q^i is matched to its nearest neighbor in target image by applying the condition of NNDR, as stated in Sect. 2.2. The generated keypoint matches are denoted as M_{1st}.

iii. D_q^i is matched to its second nearest neighbor in target image. This step will be detailed in Sect. 4.2. The generated keypoint matches are denoted as M_{2nd}.

iv. The total keypoint matches, M_f, is obtained by merging M_{1st} and M_{2nd}, i.e.,

$$M_t = M_{1st} \bigcup M_{2nd}. \tag{4}$$

v. Keypoint matches M_t is checked through to ensure the uniqueness of each keypoint match. For a particular D_q^i, if both of its nearest neighbor and second nearest neighbor are matched. Only the keypoint match from D_q^i to its nearest neighbor is preserved. After the uniqueness operation, keypoint matches are denoted as M.

vi. Keypoint matches M is refined by RANSAC [25] to obtain the final matches M_f. More details can be found in Sect. 4.3.

4.2 Keypoint Matching Beyond the Nearest Neighbor

The aim of our proposed INNDR is to improve the performance of keypoint matching over the traditional NNDR. To this end, one important side is to find more correct matches. Thus, we propose to match keypoints beyond the nearest neighbor. According to the statistics made in Sect. 3, matching to the second nearest neighbor will to some extent increase the number of correct matches. When matching the reference descriptor in query image, D_q^r, to its second nearest neighbor, the following constraint is imposed.

$$\frac{||D_q^r - D_t^{2nd}||}{||D_q^r - D_t^{3rd}||} \leq t, \tag{5}$$

where D_t^{3rd} is the third nearest neighbor of D_q^r in target image, t is a threshold of distance ratio and its value is equivalent to that in Eq. 1. To ensure a reliable matching accuracy, it is required that the second nearest neighbor should be sufficiently discriminative from the third nearest neighbor, as Eq. 5 indicates.

4.3 Why Refine Keypoint Matches?

Keypoint matching to both the nearest neighbor and the second nearest neighbor potentially lead to more number of correct matches. Meanwhile, it is vitally important to ensure a good matching performance. Let us now analyze what will happen after matching to the nearest and second nearest neighbors. First, the following denotations are given.

i. $N_{o,1st}$: the number of overall matches when matching to the nearest neighbor;

ii. $N_{c,1st}$: the number of correct matches when matching to the nearest neighbor;

iii. $N_{o,2nd}$: the number of overall matches when matching to the second nearest neighbor;

iv. $N_{c,2nd}$: the number of correct matches when matching to the second nearest neighbor.

According to the definition of matching accuracy in Eq. 9, the accuracies for matching to the nearest and second nearest neighbors are expressed as

$$accu(1st) = \frac{N_{c,1st}}{N_{o,1st}}, \tag{6}$$

and

$$accu(2nd) = \frac{N_{c,2nd}}{N_{o,2nd}}, \tag{7}$$

respectively. Intuitively, matching to the nearest neighbor generally achieves no worse accuracy as compared to the matching to the second nearest neighbor, i.e., $accu(1st) \geq accu(2nd)$. This will be validated in Sect. 5.2. This intuitive judgment leads to

$$\frac{N_{c,1st} + N_{c,2nd}}{N_{o,1st} + N_{o,2nd}} \leq \frac{N_{c,1st}}{N_{o,1st}}, \tag{8}$$

where the left of $<$ is obtained by merging those matches to the nearest and second nearest neighbors. Based on the analysis above, we are motivated to refine the keypoint matches, thus achieving better matching results. RANSAC [25][1] is used to refine keypoint matches in our implementation.

5   Performance Study

One of our main research interests is image registration based on local features. This section will evaluate INNDR against NNDR in registering images of a few benchmark datasets. As described in Sect. 4.3, refining matches by RANSAC is part of our proposed INNDR. It is interesting to see how INNDR performs when RANSAC is not inclusive. The proposed INNDR is essentially a matching strategy. For the purpose of performance evaluation, GO-SIFT and GO-IS-SIFT [18] will be used as the benchmark techniques for registering mono-modal and multi-modal images respectively.

---

[1] The matlab code of RANSAC is available at MATLAB and Octave Functions for Computer Vision and Image Processing: http://www.peterkovesi.com/matlabfns/index.html.

5.1   Evaluation Metrics and Test Datasets

The accuracy of an image registration technique depends highly on the matching accuracy. The higher the matching accuracy is, the more accurate the final registration will be [18]. Thus our proposed technique is evaluated by

$$accuracy = \frac{\#correct\ matches}{\#total\ matches} \times 100\%. \qquad (9)$$

Moreover, recall vs 1-precision [4, 26] is used for performance evaluation. The recall vs 1-precision is defined as

$$recall = \frac{\#correct\ matches}{\#correspondences} \times 100\%, \qquad (10)$$

$$1 - precision = \frac{\#incorrect\ matches}{\#total\ matches} \times 100\%, \qquad (11)$$

where #correspondences is the ground-truth number of matches. The *precision* in Eq. 11 is simply the equivalent of *accuracy* defined in Eq. 9. The ground-truths for all test image pairs are all provided. A maximum of four pixel error is considered when deciding whether a match is correct or not, which is consistent with existing literature [18, 27].

In our experiments, both mono-modal and multi-modal images are tested. In registering mono-modal images, we use the Oxford dataset [4][2] which is a benchmark dataset in image registration (Dataset 1). In this dataset, there are five different transformations: scale and rotation, viewpoint, blur, illumination and JPEG compression. This dataset contains 40 image pairs which stem from eight base images by undergoing an increasing magnitude of transformations. In registering multi-modal images, four datasets are tested. Dataset 2 is the SYM dataset[3] which includes 25 image pairs. Datasets 3 and 4 are transverse and coronal T1 vs T2 weighted MRI brain images respectively (transverse for Dataset 3 and coronal for Dataset 4). These two datasets are collected from McConnell Brain Imaging Center[4]. There are 87 and 101 image pairs in Datasets 3 and 4 respectively. In total, 253 image pairs are tested in our experiments.

5.2   Validating $accu(1st) \geq accu(2nd)$

In Sect. 4.3, we made an intuitive judgement that matching to the nearest neighbor generally achieves no worse accuracy than the matching to the second nearest neighbor, i.e., $accu(1st) \geq accu(2nd)$. Herein, this judgment is empirically validated. Figure 3 compares $accu(1st)$ against $accu(2nd)$ when registering images of Oxford dataset (Dataset 1). Overall, $accu(1st)$ is able to achieve higher accuracy as compared to $accu(2nd)$. Across all 40 image pairs in this dataset, the

---

[2] Oxford dataset: http://www.robots.ox.ac.uk/~vgg/data/data-aff.html.
[3] SYM dataset: http://www.cs.cornell.edu/projects/symfeat/.
[4] McConnell Brain Imaging Center: https://www.mcgill.ca/bic/home.

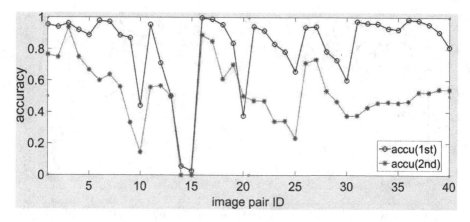

Fig. 3. Matching to the nearest neighbor and the second nearest neighbor on Oxford dataset.

average accuracy for $accu(1st)$ and $accu(2nd)$ is 81.83% and 51.93% respectively. The accuracy results on Datasets 1 to 4 show a trend similar to Fig. 3. An exceptional instance in Fig. 3 is that when registering image pair 20 $accu(2nd)$ achieves higher accuracy than $accu(1st)$. For this image pair, $accu(1st)$ achieves a 15/40 (number of correct matches/number of total matches) = 37.5% accuracy, whereas it is $1/2 = 50\%$ when it comes to $accu(2nd)$. Based on our experimental results on all four datasets, such cases happen accidentally.

5.3 Comparisons in Accuracy and Recall vs 1-Precision

Figure 4 shows matching accuracy when registering test image pairs using NNDR, INNDR without RANSAC and the proposed INNDR, respectively. Two trends can be derived from Fig. 4 as follows. First, INNDR without RANSAC generally achieves no higher accuracy as compared to NNDR and the gap between these two is quite small. Second, INNDR outperforms NNDR in general. Across all 40 pairs of Oxford dataset, the average matching accuracies for NNDR, INNDR without RANSAC, INNDR are 81.83%, 80.29% and 89.70% respectively. For all multi-modal images datasets, the average matching accuracies for NNDR, INNDR without RANSAC, INNDR are 88.53%, 86.91% and 93.47% respectively. Clearly, INNDR improves the accuracy of NNDR by a large margin.

Besides accuracy, recall vs 1-precision is utilized to compare the performance of NNDR and our proposed strategy. Four sample image pairs have been randomly selected from Datasets 1 to 4. Figure 5(a) to (d) are results of recall vs 1-precision for 2nd image pair in Dataset 1, 24th image pair in Dataset 2, 1st image pair in Dataset 3 and 7th image pair in Dataset 4 respectively. From Fig. 5 we can see that the proposed strategy obtains better performance than NNDR. Even though the INNDR without RANSAC shows advantage than NNDR. Note that, the recall vs 1-precision results for the rest image pairs show a similar

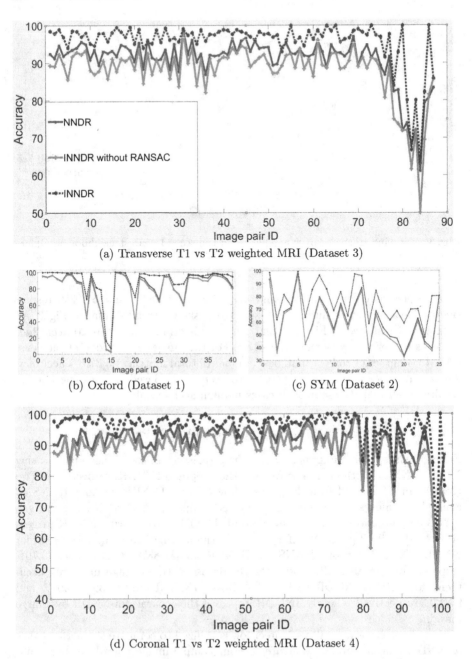

(a) Transverse T1 vs T2 weighted MRI (Dataset 3)

(b) Oxford (Dataset 1)   (c) SYM (Dataset 2)

(d) Coronal T1 vs T2 weighted MRI (Dataset 4)

Fig. 4. Matching accuracy for test datasets. In order to show the legend clearly, the accuracy results of Dataset 3 is shown first in this figure.

performance trend in most cases. Therefore, our experimental results demonstrate that the proposed INNDR outperforms NNDR in matching local image descriptors.

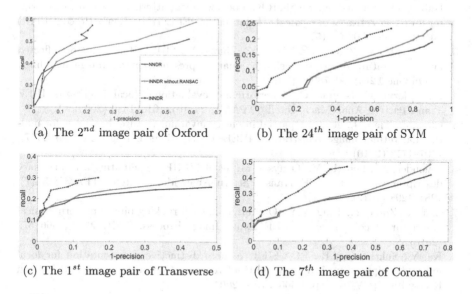

(a) The 2nd image pair of Oxford

(b) The 24th image pair of SYM

(c) The 1st image pair of Transverse

(d) The 7th image pair of Coronal

Fig. 5. Recall vs 1-precision for multimodal images.

6   Conclusions

In this paper, an improved matching strategy of local descriptors called INNDR has been proposed. The widely-used NNDR only considers the nearest neighbor and ignores other potentially correct descriptors. To attain more correct matches, the proposed INNDR takes into account both the nearest and second nearest neighbors for matching local image descriptors. Our experimental results have verified that there exist some correct matches underlying the second nearest neighbor. By making comparisons in both accuracy and recall vs 1-precision, our experimental results have shown that INNDR matching strategy outperforms NNDR in registering various types of mono-modal and multi-modal images. It is our expectation that the proposed INNDR can bring value to related research as a matching strategy of feature descriptors is essential in many fields of image processing.

References

1. Lowe, D.: Distinctive image features from scale-invariant keypoints. Int. J. Comput. Vis. (IJCV) **60**(2), 91–110 (2004)
2. Dubey, S.R., Singh, S.K., Singh, R.K.: Local wavelet pattern: a new feature descriptor for image retrieval in medical CT databases. IEEE Trans. Image Process. (TIP) **24**(12), 5892–5903 (2015)
3. Su, M., Ma, Y., Zhang, X., Wang, Y., Zhang, Y.: MBR-SIFT: a mirror reflected invariant feature descriptor using a binary representation for image matching. PLOS One **12**(5), e0178090 (2017)
4. Mikolajczyk, K., Schmid, C.: A performance evaluation of local descriptors. IEEE Trans. Pattern Anal. Mach. Intell. (TPAMI) **27**(10), 1615–1630 (2005)
5. Chen, J., Tian, J., et al.: A partial intensity invariant feature descriptor for multimodal retinal image registration. IEEE Trans. Biomed. Eng. (TBME) **57**(7), 1707–1718 (2010)
6. Calonder, M., Lepetit, V., Özuysal, M., et al.: BRIEF: computing a local binary descriptor very fast. IEEE Trans. Pattern Anal. Mach. Intell. (TPAMI) **34**(7), 1281–1298 (2012)
7. Lan, R., Zhou, Y., Tang, Y.Y.: Quaternionic local ranking binary pattern: a local descriptor of color images. IEEE Trans. Image Process. (TIP) **25**(2), 566–579 (2016)
8. Ke, Y., Sukthankar, R.: PCA-SIFT: a more distinctive representation for local image descriptors. In: International Conference on Computer Visual and Pattern Recognition (CVPR), pp. I-506–I-513 (2004)
9. Bay, H., Ess, A., Tuytelaars, T., Gool, L.V.: Speeded-up robust features (SURF). Comput. Vis. Image Underst. (CVIU) **110**(3), 346–359 (2008)
10. Morel, J., Yu, G.: ASIFT: a new framework for fully affine invariant image comparison. SIAM J. Imaging Sci. (SIIMS) **2**(2), 438–469 (2009)
11. Chen, J., Tian, J.: Real-time multi-modal rigid registration based on a novel symmetric-SIFT descriptor. Prog. Nat. Sci. (PNS) **19**, 643–651 (2009)
12. Saleem, S., Sablatnig, R.: A robust SIFT descriptor for multispectral images. IEEE Sig. Process. Lett. **21**(4), 400–403 (2014)
13. Teng, S.W., Hossain, M.T., Lu, G.: Multimodal image registration technique based on improved local feature descriptors. J. Electron. Imaging (JEI) **24**(1), 013013-1–013013-17 (2015)
14. Fan, B., Wu, F., Hu, Z.: Line matching leveraged by point correspondences. In: International Conference on Computer Vision and Pattern Recognition (CVPR), pp. 390–397 (2010)
15. Wang, Z., Fan, B., Wu, F.: Local intensity order pattern for feature description. In: International Conference on Computer Vision (ICCV), pp. 603–610 (2011)
16. Alcantarilla, P.F., Bartoli, A., Davison, A.J.: KAZE features. In: Fitzgibbon, A., Lazebnik, S., Perona, P., Sato, Y., Schmid, C. (eds.) ECCV 2012. LNCS, vol. 7577, pp. 214–227. Springer, Heidelberg (2012). https://doi.org/10.1007/978-3-642-33783-3_16
17. Lv, G.: Robust and effective techniques for multi-modal image registration. Ph.D. thesis, Monash University (2015)
18. Lv, G., Teng, S.W., Lu, G.: Enhancing SIFT-based image registration performance by building and selecting highly discriminating descriptors. Pattern Recognit. Lett. (PRL) **84**, 156–162 (2016)

19. Lv, G., Teng, S.W., Lu, G.: COREG: a corner based registration technique for multimodal images. Multimed. Tools Appl. (MTAP) **77**(10), 12607–12634 (2018)
20. Lv, G., Teng, S.W., Lu, G.: Enhancing image registration performance by incorporating distribution and spatial distance of local descriptors. Pattern Recognit. Lett. (PRL) **103**, 46–52 (2018)
21. Lv, G.: A novel correspondence selection technique for affine rigid image registration. IEEE Access **6**, 32023–32034 (2018)
22. Xiong, F., Han, X.: A 3D surface matching method using keypoint-based covariance matrix descriptors. IEEE Access **5**, 14204–14220 (2017)
23. Mendes Júnior, P.R., de Souza, R.M., Werneck, R.D.O., et al.: Nearest neighbors distance ratio open-set classifier. Mach. Learn. (ML) **106**(3), 359–386 (2017)
24. Lee, J.A., Cheng, J., et al.: A low-dimensional step pattern analysis algorithm with application to multimodal retinal image registration. In: International Conference on Computer Vision and Pattern Recognition (CVPR), pp. 1046–1053 (2015)
25. Fischler, M.A., Bolles, R.C.: Random sample consensus: a paradigm for model fitting with applications to image analysis and automated cartography. ACM Commun. **24**(6), 381–395 (1981)
26. Fan, J., Wu, Y., Wang, F., Zhang, P., Li, M.: New point matching algorithm using sparse representation of image patch feature for SAR image registration. IEEE Trans. Geosci. Remote Sens. (TGRS) **55**(3), 1498–1510 (2017)
27. Yang, G., Stewart, C.V., Sofka, M., Tsai, C.L.: Registration of challenging image pairs: initialization, estimation, and decision. IEEE Trans. Pattern Anal. Mach. Intell. (TPAMI) **29**(11), 1973–1989 (2007)

Spatial Temporal Topic Embedding: A Semantic Modeling Method for Short Text in Social Network

Congxian Yang[1], Junping Du[1(✉)], Feifei Kou[1], and Jangmyung Lee[2]

[1] Beijing Key Laboratory of Intelligent Telecommunication Software
and Multimedia, School of Computer Science,
Beijing University of Posts and Telecommunications, Beijing 100876, China
`congxianyang2016@126.com, junpingdu@126.com,`
`koufeifei000@126.com`
[2] Pusan National University, Busan, South Korea
`jmlee@pusan.ac.kr`

Abstract. Social network generates massive text data every day, which makes it important to mine its semantics. However, due to the inability to combine global semantics with local semantics, existing semantic modeling methods cannot overcome the sparseness of short texts and the ambiguity of words in different spatial-temporal environments. In this paper, we propose a semantic modeling method for social network short text, named Spatial-temporal topic embedding (STTE), which combines the spatial-temporal global context information and local context information. We first design a topic model that utilizes the text feature, time feature and location feature at the same time to generate accurate spatial-temporal global context information. Then, we employ this global information to predict an explicit topic for each word and regard the combination of each word and its assigned topic as a new pseudo word. After that, we exploit pseudo word sequence as the input of embedding vector model and finally learn the text feature which could reflect the text semantic with social network characteristics. Classification and search experiments in real-world datasets of the social network have verified that the proposed STTE has better semantic modeling ability than other baseline methods.

Keywords: Short text · Semantic modeling · Word embedding
Social network

1 Introduction

Semantic modeling is a very important task due to its extensive applications such as search, recommendation, event detection and so on. Since more and more people are accustomed to publishing and searching for what they are interested in on social networks, these platforms have carried a lot of data that need to be effectively modeled. However, texts in social networks usually are so short that will lead to data sparsity, which makes it difficult for typical semantic modeling methods to perform well on the short text of social networks. Besides of text information, there is also other attribute

Z.-H. Zhou et al. (Eds.): ICAI 2018, CCIS 888, pp. 198–210, 2018.
https://doi.org/10.1007/978-981-13-2122-1_15

information in social network data, such as the time information and location information (e.g. when and where a microblog is published). This spatial temporal information is very helpful to capture the semantics of short text, because users' interests change over time and location, which will lead to the fact that words in different spatial-temporal environment usually have different meanings. Therefore, in this paper, we aim to design a new semantic modeling method for short text in social network based on spatial-temporal information.

Typical semantic modeling methods mainly contain two kinds: topic models and word embedding. Topic models usually treat each document as a topical distribution which can get the global semantics of text from a document-level. However, traditional topic models cannot deal with the problem of data sparsity caused by the short text. Biterm feature proposed by biterm topic model (BTM) [1] has been verified is effective to generate dense semantic space. Nevertheless, these topic models just mine semantics of the short texts from a global view and do not consider the local semantics that contained in the context. Different from topic models, word embedding is another kind of popular text semantic modeling method, which extracts local text semantic information by exploiting the word co-occurrence relations in context. It represents a word as a fixed embedding vector which makes the words with similar meanings could also have similar embedding vectors. However, as the same word usually has different meanings in different spatial-temporal scenarios, the generated fixed embedding vectors cannot accurately represent the semantics of words and are unreasonable to a certain degree.

To generate more comprehensive semantics of text, some researchers have tried to mine semantics of text both from the global level and local level. The methods such as TWE [2] and NTSG [3] introduce the topic generated by topic models into the learning embedding vector. This allows the same words in different contexts represented as different features and could enhance the ability of word embedding model to distinguish polysemy to some degree. Despite this, the generated semantics is still too general, if the spatial-temporal characters can be considered in the process of the semantic modeling, the performance can be improved.

To solve the problem mentioned above, in this paper, we propose a semantic modeling method for short text in social network named spatial temporal topic embedding method (STTE). First, as people in the same place tend to focus on similar topics in a time period. We assume the documents with similar spatial temporal region have similar global context information, to utilize the global spatial temporal information, we aggregate documents with same spatial temporal region, and make the aggregated documents share similar topic distribution. Moreover, by introducing the Biterm feature into topic modeling process, we solve the semantic sparsity of input documents. After getting the topic assignment of each word in documents, we combine topics and words into pseudo-documents and use these pseudo-documents to train the embedding vectors which could reflect the precise semantic meaning of input text. The main contribution of this paper is as follows:

- We have designed a semantic modeling method for social network short text data named STTE, which could extract the accurate semantic representation of social network text.

- As far as we know, this is the first time that the global spatial-temporal context information and local context information is combined, we realize it through modeling the spatial-temporal global topic information and introducing these topics into embedding vector learning process.
- We have conducted extensive experiments on the tasks of events classification and events search to verify the semantic modeling superiority of the proposed STTE.

The rest section of this paper is organized as follows. The related work on short text semantic modeling is described in Sect. 2. We present the detail of our model in Sect. 3. The experiments settings and results analysis are shown in Sect. 4. Section 5 offers the conclusion of this paper.

2   Related Works

In order to accommodate the characteristics of social network dataset, researchers made a lot of pertinent improvements. Someone aggregate short text into long document. Zuo et al. [4] transfer the documents into a word co-occurrence network by exploiting the correlation information of words, and model the co-occurrence network instead of original documents. Yan et al. [1] generate word co-occurrence patterns by sliding window, and model word-word pairs directly. In [5], the author assumes that short text documents are generated from some pseudo documents, which contain denser semantic information, they solve short text semantic sparsity by modeling pseudo documents. Other probabilistic topic models deal with the short text semantic sparsity by making use of additional knowledge. Cai et al. [6] introduce location information into topic model, suppose documents with same location information have same topic distribution. Wang et al. [7] consider the timestamp of documents, they model topic time correlations by beta distribution.

Word embedding model is widely used in text representation. Two most prevalent embedding models Skip-Gram and continuous bag of words (CBOW) are proposed in [8], they reduce the computational complexity of traditional model, and make it possible to learn embedding vectors from large scale social network dataset. Recently, the studies about combining topic model and word embedding model become prevalent. Liu et al. [2] generate a topic for each word in corpus by using latent Dirichlet allocation, and learn embedding vectors based on both words and assigned topics. Zhang et al. [9] extend documents by regarding the topic generated by topic model as a new word, and solve the semantic sparsity by inputting the extended documents into word embedding model. Xun et al. [10] train embedding vectors on external dataset, and model topic word distribution on embedding vector space. Li et al. [11] present a model named TopicVec, different from other word embedding models, it uses embedding link function replace the latent Dirichlet allocation, which makes semantic relation encoded by cosine distance could be directly shown in embedding vector space. A word embedding model named Collaborative Language Model is proposed in [12], which exploits local and global context information by encoding both of them into word topic matrix. Besides, Liu et al. [3] improve the combination of embedding vector and assigned topic by replacing bilinear layer with a tensor layer.

3 Spatial Temporal Topic Embedding

In this section, we first give the main notations using in this paper and the general framework of the proposed model (STTE). As we utilize the global context information by assigning a topic for each word, we describe the topic inference process in Sect. 3.1. We illustrate the embedding vector learning process in Sect. 3.2.

The main purpose of STTE is to learn accurate semantic representation of social network short text, the notations of the proposed model and the corresponding explanations are described in Table 1.

Table 1. Main notations of STTE.

Symbol	Description
N_{iter}	The iteration number of Gibbs sampling
K, R	The number of topics and spatial-temporal region
M, V	The number of documents and vocabularies
N_b	The number of different word pairs
N_m^b	The number of word pairs in m^{th} document
N	The average number of words in one document
φ	The topic word distribution
θ	The topic distributions of each spatial temporal region
α, β	The hyper-parameter of region topic distribution and topic word distribution
C	The length of sliding window
L	The length of output vector

The overall framework of the STTE is shown in Fig. 1. First, by encoding the spatial-temporal region, we extract information from social network dataset. Second, we design a topic model which models the topic distribution of spatial-temporal region and word pair simultaneously, and Gibbs Sampling is used to iteratively update the distribution matrix. After the convergence of Gibbs Sampling, the word topic distribution matrix is acquired, so we can assign a global topic for each word in documents. Third, the word and the assigned topic compose a new pseudo-word, we use the pseudo-word as the input of embedding vectors training progress. By exploiting the local context information of pseudo-word sequence, we can infer the other pseudo word's appearing probability in the context of target pseudo-word, and regard this probability as the final output spatial temporal text feature (STTF).

3.1 Generative Process of Topic

A document in social network dataset can be represented by (w, t, s), w represents the word sequence of the document, t indicates the time when the document is published, s is the spatial information of the document. The following two assumptions are made about the input documents. First, each word in a word pair has same topic distribution, while word pairs are extracted by sliding window. Second, we encode spatial-temporal

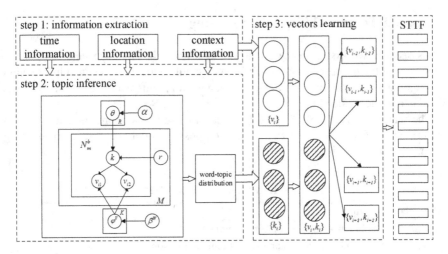

Fig. 1. STTE model.

information of input documents, both the spatial region and the temporal information is encoded by an adjustable scale, the spatial region can be countries, provinces, or cities, the temporal scale can be days, weeks, mouths, or even years. In consideration of the input documents are Chinese micro-blog and the time period of the dataset, we select provinces as the spatial scale and days as the temporal scale, and assume that the documents generated in same region on same day have same topic distribution. Based on the two hypotheses, the generative process of a social network document can be described as follow:

(1) For each encoding spatial-temporal region r, we choose a multinomial distribution θ_r as topic spatial-region distribution from Dirichlet distribution with hyper-parameter α.

(2) For each topic k, we choose a multinomial distribution φ_k as topic word distribution from Dirichlet distribution with hyper-parameter β.

(3) For each document with its spatial-temporal region r_i, we can acquire its topic distribution θ_{r_i}.

(4) For each word pair in document with its spatial-temporal region r_i. We can infer its topic k_i from θ_{r_i} and draw two words v_{i1}, v_{i2} from φ_{k_i}.

Assuming that the social network documents conform the generative process above, to obtain the topic of words, we need to infer two matrices $\{\varphi, \theta\}$. We exploit the conception mentioned in [13] and employ the algorithm named collapsed Gibbs sampling.

We first initialize the two matrices with random numbers. For each iteration process, according to the existing parameters, we infer a topic for each word pair using Eq. (1), where $\neg i$ represents the parameter value after removing i^{th} word, W represents the word pairs appearing in documents, Θ represents hyper-parameters.

$$P(k_i = k | K_{\neg i}, W, R; \Theta) \propto \frac{n_{k,\neg i} + \alpha}{\sum\limits_{k=1}^{K} n_{k,\neg i} + K\alpha} \times \frac{(n_{k,\neg i}^{v_{i1}} + \beta)(n_{k,\neg i}^{v_{i2}} + \beta)}{(\sum\limits_{v=1}^{V} n_{k,\neg i}^{v} + V\beta) \times (\sum\limits_{v=1}^{V} n_{k,\neg i}^{v} + 1 + V\beta)} \quad (1)$$

Then parameters are updated based on the new topic assigned to the word. By executing the infer progress and update progress alternately, the parameters finally reach a stable state. As a result of we model the spatial-temporal topic distribution and word topic distribution respectively, the topic distribution of each word can be calculated by Eq. (2), and the explicit topic for each word could be inferred consequently.

$$P(k_i = k | v_i, r_i) = \frac{n_{k,\neg i}^{r} + \alpha}{\sum\limits_{r=1}^{R} n_{k,\neg i}^{r} + K\alpha} \times \frac{n_{k,\neg i}^{v_i} + \beta}{\sum\limits_{v=1}^{V} n_{k,\neg i}^{v} + V\beta} \quad (2)$$

3.2 The Training Process of Embedding Vectors

After the convergence of Gibbs sampling, each word in the corpus is assigned by an explicit topic. We regard word topic pair as a pseudo-word, which used to replace the original word in documents. A three layers neural network is constructed to train embedding vectors. After one-hot encoding, the pseudo word sequence $D = \{v_1, k_1, v_2, k_2, \ldots, v_N, k_N\}$ is the input of the neural network. By moving the sliding window, we choose the input pseudo-word pair dynamically. The size of sliding window and the number of pseudo word pairs chosen in one window are adjustable, they decide the local context of v_i which is represented by C_i. The objective function of the neural network is shown in Eq. (3).

$$L(D) = \frac{1}{N} \sum_{i=1}^{N} \sum_{c \in C_i} \log \Pr(v_c | v_i) + \log \Pr(v_c | k_i) \quad (3)$$

Where $\Pr(v_c | v_i)$ represents the probability of v_c appears in the context of v_i, we use the conception in [6], employing softmax function to model it. The definition of $\Pr(v_c | v_i)$ is as follows:

$$\Pr(v_c | v_i) = \frac{\exp(v_c \cdot v_i)}{\sum\limits_{v_k \in V} \exp(v_k \cdot v_i)} \quad (4)$$

The neural network is trained by stochastic gradient descent, we update the parameters in hidden layer by exploiting backward propagation. After training process, we drop the output layer and use the corresponding parameters in the hidden layer as the embedding vector of input pseudo word.

4   Experiment and Evaluation

In this section, we evaluate the performance of STTE on social network dataset. First, we employ our model in events classification task. Second, we evaluate the quality of semantic modeling by text search on social network data.

4.1   Description of Dataset

We focus on several security events happened in recent years, and crawl data on Sina Weibo using keywords related to this events. The events we select are "Wuhan rainstorm", "Beijing severe smog" and "Tianjin explosion". The crawled data include the content of the micro-blog, the published time and user information of the micro-blog. We use the registered province of each user as the location information of each micro-blog.

In order to facilitate the experiments, we employ the following preprocessing operation. First, we remove the repeated micro-blogs in dataset. Second, we split the content of micro-blogs and then remove the stop words and words whose number of occurrences is less than five. Next, we filter out micro-blogs whose words list length is less than five or more than sixty. Finally, we utilize the spatial temporal information by encoding the spatial-temporal regions. The information of the dataset is shown in Table 2.

Table 2. Statistics of the social network short text datasets.

Dataset No.	Number of documents	Number of days	Number of provinces	Number of vocabularies
Dataset 1	24528	17	27	9970
Dataset 2	21986	15	26	11307
Dataset 3	21933	9	7	15227

4.2   Evaluation Metrics

For events classification task, precision, recall and F-measure are employed to evaluate the experiment results. The evaluation metrics on each dataset is calculated separately, and macro-averaging is used to calculate the final results. P is the precision of classification task, R represents the recall rate. β is the important weight between precision and recall, we assign it to 1, which means the precision and recall have the same importance. The definition of F-measure is shown in Eq. (5).

$$F_\beta = \frac{(\beta^2 + 1)PR}{\beta^2 P + R} \tag{5}$$

For events search experiment, we use MAP and NDCG to evaluate the performance of models. MAP means the mean average precision, is the average score for every query's AP, the AP is defined as follow:

$$AP = \frac{1}{R'} \sum_{r=1}^{R} prec(r)\delta(r) \tag{6}$$

In Eq. 6, R' represents the number of relevant results for current query statement. R represents the number of results. $prec(r)$ is the precision of top r search results. $\delta(r)$ denotes whether the rth result is related to the query statement, when r^{th} result is relevant, $\delta(r) = 1$, otherwise, $\delta(r) = 0$. The NDCG means normalized discounted cumulative gain, it assigns a ranking score to each returned result for every query statement. The definition of NDCG is shown in Eq. 7

$$\text{NDCG(n)} = C_n \sum_{i=1}^{n} \left(2^{g(i)} - 1\right) \bigg/ \log(1+i) \tag{7}$$

The C_n represents normalization factor, n denotes the total number of search results, and $g(i)$ represents the graded relevance of the ith search result.

4.3   Baseline Methods and Parameter Settings

We employ two probabilistic topic models LDA and BTM, and two word embedding models skip-gram and TWE as baseline methods.

LDA (Latent Dirichlet Allocation) [14]: LDA is a typical probabilistic topic model for long text dataset, it is usually used to extract semantic vectors by modeling the word topic distribution.

BTM (Biterm Topic Model) [2]: BTM is a probabilistic topic model for short text. It models the co-occurrence pattern of words explicitly to improve the ability for semantic modeling on short text.

Skip-gram [6]: It is a typical method for learning word embedding vectors, which extract embedding vector by predicting the possibility of other words appearing in the context of the target word.

TWE (Topical word embeddings) [7]: It is a multi-prototype embedding model and distinguishes polysemy by using latent Dirichlet allocation to generate a topic for each word.

The hyper-parameters of probabilistic topic model α and β are respectively set as 1 and 0.1, and the topics number is set as 50. The length of embedding vector is set as 300.

4.4   Events Classification

Each document in dataset is related to an event, so we evaluate the quality of extracted features by classifying the documents to explicit event. For each class of data, we randomly select 70% as training set, and 30% as test set. SVM classifiers are trained by document features learned by different models.

The length of the final embedding vector is important for semantic modeling, in order to find appropriate length of the embedding vectors, we first train the STTE using different feature length ranging from 100 to 800, the classification precision, recall and f-measure is employed to evaluate the effect of changing feature length, the average precision, recall and F-measure on the three datasets is shown in Table 3.

Table 3. Average precision recall and F-measure using different length STTF.

Length	Precision	Recall	F-measure
100	93.3	93.5	93.4
200	94.5	94.6	94.5
300	95.3	95.2	95.2
500	95.4	94.8	95.1
800	94.7	94.9	94.8

As shown in Table 3, when feature length is set to 300, the F-measure achieves a best performance, so the feature length of STTE is set to 300 when compared with other baseline method. The precision, recall rate, F-measure of every dataset is shown in Table 4, and the average scores are shown in Fig. 2.

Table 4. Precision recall and F-measure on each dataset.

Algorithm	Dataset 1			Dataset 2			Dataset 3		
	Precision	Recall	F-measure	Precision	Recall	F-measure	Precision	Recall	F-measure
LDA	83.6	86.2	84.9	81.2	81.8	81.5	82.3	80.8	82.5
BTM	90.0	90.6	90.2	84.2	87.0	85.6	86.6	82.9	84.7
Skip-Gram	85.5	85.7	85.6	84.3	89.1	86.6	86.2	81.0	83.5
TWE	92.1	92.1	92.1	92.2	90.3	91.2	95.6	97.9	96.8
STTE	95.8	95.4	95.6	93.7	99.4	96.4	96.3	90.8	93.5

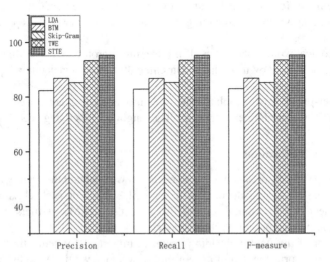

Fig. 2. Average precision recall and F-measure on three datasets.

We can get the following conclusions from the scores of evaluation metrics: LDA is a probabilistic topic model designed for long text, which just models the topic distribution for word. When employing LDA on modeling social network short text semantic, its ability to distinguish different events is always poor. The document vectors extracted by BTM perform much better than LDA in events classification. Although BTM is also a probabilistic model, it can solve the semantic sparsity of social network text to some degree due to it models the words co-occurrence pattern explicitly. Skip-Gram also performs relatively poor than our STTE, which is because of the training process of Skip-Gram always need large external dataset. But as a result of exploiting local context information, the precision of classification using Skip-Gram features is better than LDA. The embedding vectors learned by TWE have a better performance on events classification than Skip-Gram. TWE assigns a topic to each word in documents, this makes the final embedding vectors generated by TWE contain the global context information, and get better classification performance. Our model STTE shows the best performance on every dataset. This is because it exploits spatial-temporal global context information by assigning topics for every word appears in documents, and learns embedding vectors by using local context information.

4.5 Events Search

In this section, we employ events search experiment to evaluate the quality of text features extracted by STTE and baseline methods. For each class in dataset, we select 100 documents as query statements. The similarity of different text features is calculated by cosine distance defined in Eq. 8.

$$\text{similarity}(A,B) = \frac{A \cdot B}{\|A\|\|B\|} \tag{8}$$

We choose top 10, 20, 30 search results to compute the MAP scores, the MAP@R scores on each database are shown in Fig. 3, the average scores of MAP@R is shown in Fig. 4. In order to calculate the NDCG scores, we invited ten volunteers to evaluate the top 15 search results for each query statement, using five level to score the results. The average score of NDCG@n is shown in Table 5.

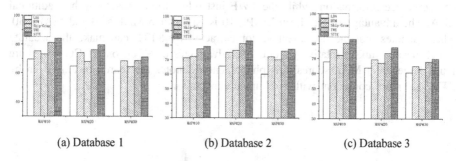

(a) Database 1   (b) Database 2   (c) Database 3

Fig. 3. MAP@R scores on each database.

Table 5. Average NDCG scores.

Algorithm	NDCG@5	NDCG@10	NDCG@15
LDA	60.3	63.5	59.5
BTM	72.4	75.6	73.3
Skip-Gram	68.8	71.2	66.9
TWE	77.6	80.2	76.3
STTE	85.6	86.2	83.7

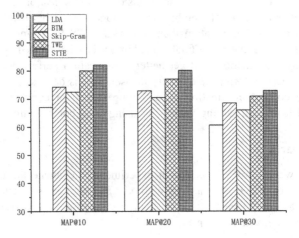

Fig. 4. Average MAP scores.

As shown in Table 5 and Fig. 4, we can get the following observations. The models (TWE and STTE) introducing topics into embedding vectors training process have a better performance, this is because the topic assigned to the word contains additional global context information, which can enhance the ability of distinguishing polysemy, making the search results more similar in semantic space. The performance of STTE is better than TWE, this is because of the topic assigned by STTE is more accurate. STTE infers topics by modeling spatial-temporal topic distribution and word co-occurrence information, while the TWE just infers topics based on the traditional LDA. The advantage of STTE on MAP@10 is larger than MAP@20 and MAP@30, which denotes that the document vector extracted by STTE can make the accurate search results rank in relatively front. The advantages of STTE on NDCG scores are much larger than MAP scores, this phenomenon denotes the search results given by STTE are more consistent with the human expectation.

5 Conclusion

In this paper, we propose a semantic modeling method for social network short text named STTE based on spatial-temporal global information and local context information. There are several advantages of STTE. First, it can solve semantic sparsity of social network data by aggregating the documents published in the same spatial-temporal region, and modeling the topic distribution of spatial-temporal region and word pairs. Second, by using the combination of topics generated by global context information and the original words, STTE generates embedding vectors of social network short text, which contain both global and local context information. The experiments results of classification and search show that compared with the start-of-art text semantic modeling algorithms, the proposed STTE achieves better performance on social network text data, and has a promising application prospect.

Acknowledgements. This work is supported by the National Natural Science Foundation of China (No. 61320106006, No. 61532006, No. 61772083).

References

1. Yan, X., Guo, J., Lan, Y., et al.: A biterm topic model for short texts. In: Proceedings of the 22nd International Conference on World Wide Web, pp. 1445–1456. ACM (2013)
2. Liu, Y., Liu, Z., Chua, T.S., et al.: Topical word embeddings. In: AAAI, pp. 2418–2424 (2015)
3. Liu, P., Qiu, X., Huang, X.: Learning context-sensitive word embeddings with neural tensor skip-gram model. In: International Conference on Artificial Intelligence, pp. 1284–1290. AAAI Press (2015)
4. Zuo, Y., Zhao, J., Xu, K.: Word network topic model: a simple but general solution for short and imbalanced texts. Knowl. Inf. Syst. **48**(2), 379–398 (2016)
5. Zuo, Y., Wu, J., Zhang, H., et al.: Topic modeling of short texts: a pseudo-document view. In: ACM SIGKDD International Conference on Knowledge Discovery and Data Mining, pp. 2105–2114. ACM (2016)
6. Cai, H., Yang, Y., Li, X., et al.: What are popular: exploring Twitter features for event detection, tracking and visualization. In: ACM International Conference on Multimedia, pp. 89–98. ACM (2015)
7. Kou, F., Du, J., Lin, Z., et al.: A semantic modeling method for social network short text based on spatial and temporal characteristics. J. Comput. Sci. (2017)
8. Mikolov, T., Chen, K., Corrado, G., et al.: Efficient estimation of word representations in vector space. Comput. Sci. (2013)
9. Zhang, H., Zhong, G.: Improving short text classification by learning vector representations of both words and hidden topics. Knowl.-Based Syst. **102**, 76–86 (2016)
10. Xun, G., Li, Y., Zhao, W.X., et al.: A correlated topic model using word embeddings. In: Twenty-Sixth International Joint Conference on Artificial Intelligence, pp. 4207–4213 (2017)
11. Li, S., Chua, T.S., Zhu, J., et al.: Generative topic embedding: a continuous representation of documents. In: Meeting of the Association for Computational Linguistics, pp. 666–675 (2016)

12. Xun, G., Li, Y., Gao, J., et al.: Collaboratively improving topic discovery and word embeddings by coordinating global and local contexts. In: The ACM SIGKDD International Conference, pp. 535–543. ACM (2017)

13. Li, A.Q., Ahmed, A., Ravi, S., et al.: Reducing the sampling complexity of topic models. In: ACM (2014)

14. Blei, D.M., Ng, A.Y., Jordan, M.I.: Latent Dirichlet allocation. J. Mach. Learn. Res. 3(Jan), 993–1022 (2003)

A Pulse Wave Based Blood Pressure Monitoring and Analysis Algorithm

Yue Liu[1,2], Xin Sun[3], Yongmei Sun[4(✉)], Kuixing Zhang[1,2],
Yanfei Hong[1,2], and Benzheng Wei[1,2(✉)]

[1] College of Science and Technology,
Shandong University of Traditional Chinese Medicine, Jinan 250355, China
[2] Computational Medicine Lab,
Shandong University of Traditional Chinese Medicine, Jinan 250355, China
wbz99@sina.com
[3] Department of Medical Engineering,
Shandong Provincial Hospital Affiliated to Shandong University,
Jinan 250021, China
[4] Shandong Provincial Third Hospital, Jinan 250023, China
15069027902@163.com

Abstract. Pulse wave reflects the heartbeat, pulse rhythm and the dynamic state of angiocarpy, but the feature extraction and auto-identification of hypertension based on pulse wave are difficult and crucial for monitoring and analyzing pulse signal. To address this key problem, an algorithm framework is proposed for auto-identification of hypertension, which named HHT-ELM analysis and recognition algorithm. The proposed algorithm framework includes preprocessing module, feature extraction module and signal recognition module. The first module adopts the low-pass filter and the morphological filter to denoise the original signal from the sensor. The Hilbert-Huang transform (HHT) is utilized to extract characteristics, and the empirical mode decomposition (EMD) is the main component of HHT, which is used to disassemble the pulse wave to acquire multiple intrinsic modal functions (IMF). Then the Hilbert transformation of each IMF yields the Hilbert spectra and the marginal spectra to extract the feature in the second module. Finally, the extreme learning machine (ELM) is employed to classify features, so the automatic recognition can be achieved in the last module. The study results show that the algorithm can diagnose hypertension with an accuracy rate of over 93%. Therefore, it provides a novel method of automatic pulse signal process and analysis for the clinical diagnosis and portable monitoring of hypertension.

Keywords: Pulse wave · Hypertension · Hilbert-Huang transform
Extreme learning machine

1 Introduction

Pulse wave is crucial physiological parameter to human body, which has great significance for the diagnosis of angiocardiopathy [1]. Most researches have revealed that the common cardiovascular diseases can be reflected from the pulse wave, such as

© Springer Nature Singapore Pte Ltd. 2018
Z.-H. Zhou et al. (Eds.): ICAI 2018, CCIS 888, pp. 211–221, 2018.
https://doi.org/10.1007/978-981-13-2122-1_16

hypertension, coronary heart disease, angina pectoris, etc. Physicians incessantly acquire blood pressure, blood oxygen and heart rate to understand the continuously monitored patient's physical condition in ICU and inpatient department. Moreover, with the development of information and mobile technology, it can be achieved easily to continuously monitor blood pressure using mobile phones and portable devices. However, the diagnostic precision of hypertension based on pulse signal is sometimes inaccurate either in clinical areas or by portable devices. Therefore, this paper proposes an HHT-ELM analysis and recognition algorithm to address this problem, which can conveniently and accurately diagnose hypertension by pulse signals.

2 Related Works

The Korotkoff auscultation had been widely utilized as the standard of blood pressure measurement in the clinical field. However, noninvasive blood pressure measurement technology had gradually replaced it with the development of sensor technology, which could be divided into two types from monitoring mode: intermittence and continuity [2]. Intermittent blood pressure measurement method included Korotkoff auscultation method, oscillation method and ultrasonic method; and the continuity method mainly included volume compensation method, tension measurement method, pulse blood volume detection method, Gu et al. [3]. The continuity method had become the main monitoring method for its convenient and real-time performance.

Pulse wave acquisition mainly employs the electrode or sensor to contact the human body's detection site, and then the pulse signal is obtained by magnifying and filtering the.analog signal. There were various methods for pulse signals processing and analysis in recent years. Wavelet transform (WT) and HHT are the two most common methods for pulse signals processing. WT is an improved method based on Fourier transform, which extracts features by decomposing and reconstructing signals. HHT was proposed by Dr. Huang and his team in 1998. This team made an innovative study of the theory of signal analysis and introduced the concept of IMF, Hilbert spectra and marginal spectra [4]. HHT and WT can be used for non-linear and non-stationary signal processing, but HHT's basic function is adaptable, so it doesn't need to be adjusted according to different signals [5]. The results of HHT are IMF, Hilbert spectrum and marginal spectrum, which are plain and their feature points are conspicuous, so this module utilizes HHT as signal feature extraction algorithm. The HHT algorithm had been widely employed in medical field for the past few years. Li adopted HHT algorithm to analyze the effect of stent implantation on pulse wave in patients with coronary heart disease [6]. Hong also utilized the same method to analyze the pulse signals, and his team used EMD to break the pulse signals into multiple IMF signals and described the biological significance of each modality [7]. Followed by the above researches, the HHT algorithm is employed as the prophase algorithm of pulse signals processing.

Adaboost, SVM and ELM are commonly classification method in signal processing field. The Adaboost algorithm can classify all samples and generate a weak classifier by automatically finding the optimal features, which minimize the number of misclassified samples. Then the training process is iterated to form a weak classifier set and a strong classifier is generated in a cascaded manner [8]. SVM was proposed by Vapnick on the basis of statistical learning theory, which could analyze high-dimensional and nonlinear characteristics of samples. ELM is a new learning algorithm of single hidden layer feed forward neural network, which was proposed by Huang in 2004 [9]. According to the characteristics of pulse signals, ELM algorithm is selected as the classification recognition algorithm of this module. This algorithm was widely used in image processing because of its higher stability, stronger generalization ability, faster learning speed and less restrictive conditions [10–14].

Most of the above studies focused on the pulse wave feature extraction rather than the auto-identification of pulse wave related diseases. Therefore, an HHT-ELM analysis and recognition algorithm was proposed to achieve continuous monitoring of the pulse wave, which used pulse wave to automatically identify hypertension and could assist clinical diagnosis.

3 HHT-ELM Analysis and Recognition Algorithm

As shown in Fig. 1, the proposed algorithm framework consists of three modules: preprocessing module, feature extraction module and signal recognition module. The first module is used to suppress the noise from the pulse signals, after this preprocessing step, the pulse signal is decomposed and the specific characteristics are extracted in the second module. And the last module classifies the features and automatically identifies the pulse signals of hypertension. The algorithm steps can be described as follows:

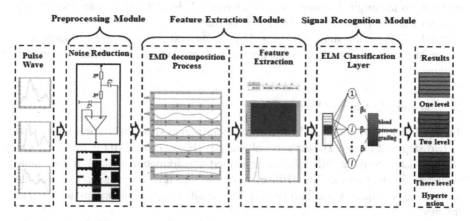

Fig. 1. HHT-ELM analysis and recognition algorithm frame diagram

Step 1: The acquired pulse signals utilize the low-pass filter to denoise the high-frequency noise, and then the achieved signals adopt the morphological filter to eliminate baseline drift.

Step 2: The preprocessing pulse signals are decomposed by EMD to gain four IMF components and one residual component.

Step 3: The energy of each IMF component is calculated by the algorithm and the corresponding energy distribution table is got.

Step 4: The Hilbert spectra and marginal spectra are acquired by Hilbert transformation of each IMF component.

Step 5: The IMF components, energy distribution tables, Hilbert spectra and marginal spectra are synthetically analyzed to find out the characteristic points.

Step 6: The extracted features are classified by ELM to provide the basis for auto-identification of hypertension.

Followed those above steps, the patient's pulse signals can be automatically identified and the hypertension can be quickly diagnosed. The details of each module are described as follows.

3.1   The Preprocessing Module

Pulse signals are susceptible to various background noises in the process of collection, including environmental disturbances, human biological noises and baseline drift. Considering the characteristics of the signal, the low-pass filter is utilized to undo the high frequency noises, and then the baseline drift is restrained by the morphological filter [15]. The latter filter can eliminate the noises by the opening and closing operator of the original signals. Its expression can be described as follows:

$$x(t) = f(t) - \frac{[f(t) \circ l_1 \bullet l_1 + f(t) \bullet l_1 \circ l_1]}{2} \quad (1)$$

where $f(t)$ is the original signals and l_1 is a structural element. "$\circ$" denotes the opening operator and "$\bullet$" is the sign of the closing operator.

Each original pulse signal is extracted by this method as a 1 * 116 one-dimensional vector, and the next feature extraction module will decompose these vectors and extract features.

3.2   Feature Extraction Module

The HHT algorithm is adopted in the feature extraction module, and its process consists of two steps: EMD and Hilbert spectral analysis (HSA) [16]. EMD is the decomposition of data into finite IMF. HSA is the Hilbert transformation of the IMF to acquire an energy distribution spectrum (Hilbert spectrum) on a time-frequency plane to obtain instantaneous frequency and energy [17]. The detailed steps of algorithm are listed as follows:

Step 1: All local extreme points of time series $x(t)$ are determined, and then all the maximum and minimum points are connected with spline curves and the upper and lower envelope lines of $x(t)$ are get, which is called $m(t)$.

Step 2: A new signal $h_1(t)$ is obtained by removing the envelope mean $m(t)$ from the original signal $x(t)$.

$$h_1(t) = x(t) - m(t) \tag{2}$$

There are two necessary conditions for judging $h_1(t)$ to satisfy the IMF, one is that the number of extreme points is equal or less to the number of zero crossing points in the entire data sequence, and the other is that the mean value of the local upper and lower envelopes of the signal is zero. If the above conditions can't be met, the $h_1(t)$ is repeated as an input signal until the $h_1(t)$ satisfies the IMF's terms, which is called $c_1(t)$.

Step 3: The original signal $x(t)$ removes the first IMF $c_1(t)$ to get the sequence $x_1(t) = x(t) - c_1(t)$. The $x_1(t)$ is taken as an input sequence and the above steps are repeated to get next IMF. When the decomposed IMF is negligible or monotone and it can't continue to decompose the IMF function, the whole process is stopped. The original signal $x(t)$ can be expressed as the sum of all IMF functions and a residual function $r(t)$:

$$x(t) = \sum_{i=1}^{n} c_i(t) + r(t) \quad (i = 1, 2, 3 \cdots n) \tag{3}$$

Step 4: The HHT is carried out for the each IMF function because it can get the physical meaning of the instantaneous frequency.

$$x(t) = Re \sum_{i=1}^{n} c_i(t) e^{j\varphi t} = Re \sum_{i=1}^{n} c_i(t) e^{j\int \omega_i(t) d(t)} \tag{4}$$

Since the remainder $r(t)$ is a small constant or a monotone function and it can be omitted, so that the Hilbert spectral formula $H(\omega, t)$ of $x(t)$ is defined as

$$H(\omega, t) = \begin{cases} Re \sum_{i=1}^{n} c_i(t) e^{j\int \omega_i(t) d(t)} & \omega_i(t) = \omega \\ 0 & \omega_i(t) \neq \omega \end{cases} \tag{5}$$

The energy of each IMF E_i can be described as follows:

$$E_i = \int_{-\infty}^{+\infty} |c_i(t)|^2 dt \tag{6}$$

And it is summed as

$$E = \sum_{i=1}^{n} E_i \tag{7}$$

Step 5: The normalized process is carried out to avoid the calculation problem caused by excessive energy amplitude [7]. The normalized formula T_i is described as follows:

$$T_i = \frac{E_i}{E} \tag{8}$$

Through signal decomposition and analysis, the algorithm identifies specific features and extracts the signal parameters of the corresponding frequency band. The 1 * 36 eigenvector is achieved from each signal as the input weight vector for the next signal recognition module.

3.3 Signal Recognition Module

The ELM algorithm is adopted in the signal recognition module, and the mathematical expression of the ELM training model can be described as [12]:

$$\sum_{i=1}^{M} \beta_i g(\delta_i x_i + b_i) = o_j (j = 1, 2, 3, \cdots \cdots N) \tag{9}$$

$\delta_i = [E_i, x(t)]$ is the input weight vector, where the input is the characteristics acquired by HHT: the IMF and the energy; β_i is the output weight vector; o_j represents the grid output value; b_i represents a hidden layer neuron threshold. The cost function G of the ELM can be expressed as:

$$G(S, \beta) = \sum_{j=1}^{N} \|o_j - t_j\| \tag{10}$$

where $S = (\delta_i, b_i)$ includes the input weights of neural networks and the threshold values of hidden layer neurons. By the minimizing cost function $\min(G(S, \beta))$, ELM seeks the optimal output weight β, that is the error between the minimizing network output value and the corresponding actual measured value. In practical application, if HHT is not singular, β can be obtained by formula (11):

$$\beta = H^- O = H^- (HH^T)^{-1} O = H^T (1/C + HH^T) O \tag{11}$$

The H is the hidden layer matrix of neural network; H^- is the generalized inverse matrix of H-matrix; O is a vector of predicted target value [10].

The feature vector sets are divided into five parts, then the four sets are selected as training set and the other one set is as test set to complete the cross-validation. So the classification results can be achieved by this method. The number of hidden neurons can lead the problems of overfitting and underfitting based on the experimental scale. Therefore, the number of hidden neurons is set as 50 to conquer this problem in the ELM, this empirical parameters is determined by a lot of experiments.

4   Experimental Results and Analysis

4.1   Data Sources

The data set of pulse signals was collected from hypertensive patients and normal people by employing HK-2000C pulse transducer, and the most stable pulse wave was extracted from each person. In order to ensure the validity of the pulse signals, all data was acquired in the same time frame. 72 cases of hypertensive were collected, which were programmed into hypertension experimental group, and 70 normal people were catalogued into normal control group. The pulse signals of the hypertension group were provided by the Xinqiaodong community hospital in Yantai, Shandong province, and the data of the control group was offered by the workers of Shandong Hengmin medical corporation.

The criteria for the determination of hypertension patients are according to "Guidelines for prevention and treatment of hypertension": Normal blood pressure range is 90–140 mmHg systolic pressure and 60–90 mmHg diastolic pressure. If the patient has a higher blood pressure than normal systolic pressure or diastolic pressure, he/she can be determined as a hypertensive [18]. The data set acquired is strictly in accordance with this standard.

4.2 Data Preprocessing

After the processing of the low-pass filter and morphological filter, the original signal and the preprocessing signal diagrams are shown in Figs. 2 and 3:

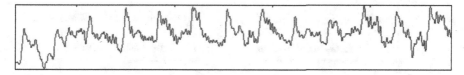

Fig. 2. Schematic diagram of the original pulse signals

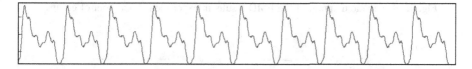

Fig. 3. Schematic diagram of the pulse signal after pretreatment

4.3 Result Analysis

The proposed HHT-ELM algorithm is implemented in MatlabR2014a, which first decomposes the pulse signals to get IMF, energy distribution, Hilbert spectra and marginal spectra. In Fig. 4, we show the EMD results of the hypertensive experimental group and the normal control group. It can be found that the imf1 of the experimental group fluctuates slightly from the baseline of 0, while the imf1 of the control group shows a spike near 20 units of time. The amplitude of imf3 in the experimental group is large, which shows that there is a large peak in the low frequency band of the hypertensive patients' pulse signals, while the normal people are stable. From Table 1, the energy of the experimental group mainly concentrates in imf2, imf3 and imf4, while the control group mainly centralizes in imf3 and imf4. In the experimental group, the energy of the imf4 is supreme and the imf2 and imf3 also contain a lot of energy, but the control group concentrates in the imf3. The proposed method also extracts features in the frequency domain, as shown in Fig. 6. Both the experimental group and the control group have energy peaks at 5 Hz, while the experimental group has a prominent peak in the 5–10 Hz. This shows that the energy accumulation of hypertensive patients is much greater than that of normal people in this interval. At the same time, we also extract the characteristics of pulse signals by WT. The extraction results are shown in Fig. 5. The cd1 and cd2 waveforms of the control group are relatively stable, while the experimental group has larger amplitude in these two frequency bands. The cd3 of the control group shows a clear peak near 5 units of time, while the peak of the experimental group is not obvious. The feature dimensions extracted by HHT and WT are the same and these feature vectors are used for classification recognition.

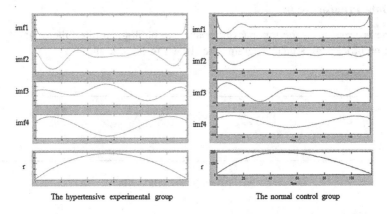

The hypertensive experimental group   The normal control group

Fig. 4. Schematic diagram of the EMD results in experimental and control groups

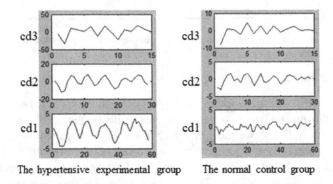

The hypertensive experimental group   The normal control group

Fig. 5. Schematic diagram of the WT results in experimental and control groups

Table 1. The IMF energy distribution in experimental and control groups

IMF	Hypertensive experimental group	Normal control group
imf1	26.0006	28.3212
imf2	1.0306×10^4	898.5382
imf3	1.3366×10^3	1.0575×10^6
imf4	1.3927×10^6	6.6834×10^3

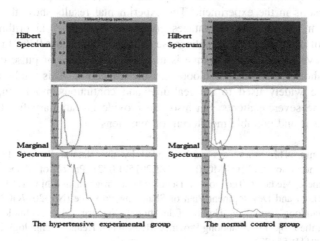

Fig. 6. Spectrum diagram in experimental and control groups

The characteristic data mentioned above with obvious specificity as input values are classified by ELM. The SVM and Adaboost classifiers are utilized in this paper, and their experimental results are compared with the ELM's. Classification experiments take cross-validation to get the average. The result of classification accuracy is shown in Table 2. The result shows that the diagnostic accuracy of auto-identification of hypertension can reach more than 72% after learning and classifying by Adaboost, SVM and ELM. This indicates that the characteristic signals extracted are obviously diverse. Among them, the diagnostic accuracy of HHT-ELM is more than 93%. Compared with the results, the superiority of HHT-ELM algorithm is demonstrated clearly.

Table 2. Comparison of diagnostic accuracy of different algorithms

Algorithm	Accuracy of recognition/%
WT-Adaboost	72.76
HHT-Adaboost	79.82
WT-SVM	77.09
HHT-SVM	88.00
WT-ELM	80.17
HHT-ELM	93.67

5   Conclusion

HHT is a popular signal processing methods and has been widely applied for pulse signal analysis. However, traditional HHT algorithm can't well meet the needs of pulse signal analysis directly. Therefore, an HHT-ELM analysis and recognition algorithm is proposed in this paper, which combines ELM with HHT to extract the more representative features of the pulse signal. The two feature extraction methods (HHT and WT) and three classifiers (Adaboost, SVM and ELM) are adopted to compare the recognition results in the experiment. The experimental results show that the HHT-ELM algorithm has obvious advantages compared with other algorithms and the accuracy rate of using pulse wave to diagnose hypertension reaches 93%. In the clinical field, pulse wave acquisition technique is mature and the time of pulse collection is significantly shorter than that of blood pressure collection. This study is not only expected to be widely used to the real-time and continuous monitoring of blood pressure for the severe patients, but also can provide an auxiliary function for the clinical diagnosis and portable monitoring of hypertension.

Acknowledgments. This work was partly funded by the Natural Science Foundation of Shandong Province (No. ZR2015FM010, No. ZR2015FL022), Project of Shandong Province Traditional Chinese Medicine Technology Development Program in China (2015-026, 2017-001), Key Research and Development Plan of Shandong Province (No. 2017GGX10139), the Project of Shandong Province Higher Educational Science and Technology Program (No. J15LN20), the Project of Shandong Province Medical and Health Technology Development Program (No. 2016WS0577).

References

1. Wang, M., Sun, X., Wei, B.Z.: Design and development of detection equipment based on the "O Curve" method for blood pressure simulator. China Med. Devices **33**(01), 39–42 (2018)
2. Wei, Y.B., Wei, Z., Huang, W.C., et al.: Non-invasive human extremely weak pulse wave measurement based on a high-precision laser self-mixing interferometer. Optoelectron. Lett. **13**(02), 143–146 (2017)
3. Gu, Y.X., Yang, T., Bao, K., et al.: Study on multi-mode calculation model in non-invasive blood pressure measurement by pulse wave velocity method. Chin. J. Biomed. Eng. **06**(35), 691–698 (2016)
4. Huang, N.E., Shen, Z., Long, S.R., et al.: The empirical mode decomposition and the Hilbert spectrum for nonlinear and non-stationary time series analysis. Roy. Soc. **454**, 903–995 (1998)
5. Huang, N.E., Hu, K., Yang, A.C., et al.: On Holo-Hilbert spectral analysis: a full informational spectral representation for nonlinear and non-stationary data. Philos. Trans. Roy. Soc. A Math. Phys. Eng. Sci. **374**(2065), 20150206 (2016)
6. Li, F.F., Sun, R., Xue, S., et al.: Pulse signal analysis of patients with coronary heart diseases Using Hilbert-Huang transformation and time-domain method. Chin. J. Integr. Med. **21**(05), 355–360 (2015)
7. Hong, G.: Refers to analysis of finger pulse feature and its application research. Shandong University of Traditional Chinese Medicine, Shandong (2016)

8. Zang, P.P., Wei, B.Z., Zhang, Q., et al.: Optic disc detection based on Ada Boost. J. Univ. Jinan (Sci. Technol.) **30**(3), 220–225 (2016)
9. Huang, G.B., Zhu, Q.Y., Siew, C.K.: Extreme learning machine: a new learning scheme of feedforward neural networks. 2004 IEEE International Joint Conference on Neural Networks, vol. 2, pp. 985–990. IEEE (2004)
10. Toh, K.-A.: Deterministic neural classification. Neural Comput. **20**(6), 1565–1595 (2008)
11. Wei, B.Z., Zhao, Z.M.: A sub-pixel edge detection algorithm based on Zernike moments. Imaging Sci. J. **61**(5), 436–446 (2013)
12. Huang, G.B., Ding, X., Zhou, H.: Optimization method based extreme learning machine for classification. Neurocomputing **74**, 155–163 (2010)
13. He, Y.L., Geng, Z.Q., Zhu, Q.X.: Positive and negative correlation input attributes oriented subnets based double parallel extreme learning machine (PNIAOS-DPELM) and its application to monitoring chemical processes in steady state. Neurocomputing **165**(1), 171–181 (2015)
14. Iosifidis, A., Tefas, A., Pitas, I.: Graph embedded extreme learning machine. IEEE Trans. Cybern. **46**(1), 311–324 (2016)
15. Wei, B.Z., Zhao, Z.M., Peng, X.A.: Novel method of medical image registration based on feature point mutual information and IPOS algorithm. J. Comput. Inf. Syst. **7**(2), 559–567 (2010)
16. Zhang, K.X., Luo, F.F., Wei, B.Z.: Analysis and research on spectrum signal of mice blood based on Hilbert-Huang transform. China Med. Equip. **11**(9), 12–14 (2014)
17. Huang, N.E., Shen, Z., Long, S.R.: A new view of nonlinear water waves: the Hilbert spectrum. Annu. Rev. Fluid Mech. **31**, 417 (1999)
18. Afsar, B., Elsurer, R.: The relationship between magnesium and ambulatory blood pressure, augmentation index, pulse wave velocity, total peripheral resistance, and cardiac output in essential hypertensive patients. J. Am. Soc. Hypertens. **8**(1), 28–35 (2014)

An Improved Collaborative Filtering Algorithm Based on Bhattacharyya Coefficient and LDA Topic Model

Chunxia Zhang and Ming Yang[(✉)]

School of Computer Science and Technology, Nanjing Normal University,
Nanjing 210023, China
1608195294@qq.com, myang@njnu.edu.cn

Abstract. Collaborative filtering (CF) is the most successful method used in designing recommendation systems, which includes the neighbor-based method and the model-based method. Traditional neighbor-based method calculates similarity only based on the rating matrix, but the rating matrix is very sparse. Therefore, to address the problem of sparsity, we proposed an improved collaborative filtering algorithm unified Bhattacharyya coefficient and LDA topic model (UBL-CF). UBL-CF utilized the LDA topic model to mine potential topic information in the tag set and embed the underlying topic information into the progress of the calculation of similarity. Meanwhile, it introduces Bhattacharyya coefficient to alleviate the data sparsity without common ratings. Experimental results show that our method has better prediction in accuracy.

Keywords: Recommendation system · Similarity · LDA
Bhattacharyya coefficient

1 Introduction

With the popularity of the Internet and the frantic growth of information, personal processing capabilities have become overwhelmed. This phenomenon is called information overload [1]. How to obtain useful information from the complex information world and solve the problem of information overload has become a challenging task.

Recommendation system [2, 3] is an important branch of the research field of personalized services. Through the analysis of users' historical behavior, the users' hidden preferences and the underlying deep relationships between users and items are discovered, helping users find items they may be interested in (such as information, online products, services, etc.). In the recommendation system, the recommendation algorithm is an important component that supports its operation. At present, the main recommendation techniques include content-based recommendation [4, 5], collaborative filtering-based recommendation [6–9], knowledge-based recommendation [10, 11], and hybrid recommendation [12, 13]. Among them, collaborative filtering recommendation algorithm is widely used because it is simple and easy to implement.

The rating prediction is an important issue in the recommendation system. Researchers have done a lot of related work on collaborative filtering algorithms based

© Springer Nature Singapore Pte Ltd. 2018
Z.-H. Zhou et al. (Eds.): ICAI 2018, CCIS 888, pp. 222–232, 2018.
https://doi.org/10.1007/978-981-13-2122-1_17

on rating predictions, and have made great progress. In the neighbor-based collaborative filtering algorithm, the similarity calculation between users or items is the key step. With a more accurate similarity calculation method, a more effective neighbor set can be found and a more accurate recommendation can be made. With the rapid development of Web 2.0, more and more users have begun to personalize their expressions on social networks. Today, users' different tag information freely express their own ideas. Therefore, a large amount of user preference information is implied in the social tag. If it is applied to the recommendation process, it will greatly help the recommendation result.

2 Related Work

Collaborative filtering (CF) is the most successful method used in designing recommendation systems. CF does not require collecting large amounts of user's or item's information, nor does it require any domain knowledge. It is mainly divided into neighbor-based collaborative filtering [6, 7] and model-based collaborative filtering [8, 14, 15]. Neighbor-based collaborative filtering calculates the similarity between users (or items) from the rating matrix, selects neighbor users with similar interests for the current user, and then make recommendations to the current user according to the neighbor user's known rating. Neighbor-based collaborative filtering includes user-based collaborative filtering and item-based collaborative filtering. Different from neighbor-based collaborative filtering, model-based collaborative filtering applied machine learning, statistics, data mining and other methods to model the user's preferences by analyzing the user's previous purchase behavior, get the current user's preference model and then make recommendations. However, data sparsity has always been a difficult gap for collaborative filtering recommendation algorithms. There is often no way to recommend new users and new items without information. To solve these problems, many scholars have put forward a lot of methods on how to improve the recommendation technology. The literature [16] proposes a model that combines collaborative filtering and user information to deal with cold start problems based on the probabilistic model analysis. The literature [17] proposes a method based on demographic statistics and collaborative filtering to enhance the effectiveness of recommendations by using demographic information. The literature [18] combines the decision tree and collaborative filtering algorithm to recommend new users, mainly based on the user-item matrix, clustering users, using the clustering results and the user's demographic information to create a decision tree to obtain new relationship between users and existing users. The literature [19] introduces credit ratings to filter out important ratings and greatly reduces the dimensions of the user-item matrix. Literature [20] proposes a collaborative filtering algorithm based on singular value decomposition, which uses a low rank approximation to remove data noise caused by unstable users to improve the accuracy of the recommendation. In recent years, social tagging systems have become more and more popular on the Internet. Tag-based recommendation methods have emerged. Researchers can improve the quality of recommendations by digging out useful information from the tag information that exists in the system. The literature [21] makes full use of tag information to find neighbor sets for users and items, alleviates data sparseness problems and improves the accuracy of recommendations; the literature [22] builds a tag-keyword relationship matrix based on a user-item rating matrix and introduces it into the matrix decomposition algorithm to mitigate the cold start problem. With the rapid development of the topic model in the field of natural processing, many scholars have applied the Latent Dirichlet Allocation (LDA) topic model to the recommendation system to obtain extra information by processing the text. Literature [23] proposes a collaborative filtering algorithm that combines LDA topic models to mine user interest and improve the quality of recommendations. Literature [24] uses the LDA topic model to learn from the user's priorities for item rating.

3   Our Model

3.1   Bhattacharyya Coefficient

The Bhattacharyya coefficient (BC) is commonly used to calculate the similarity between the color probability distributions of the reference template and the candidate template. It is popular in signal processing and pattern recognition research.

The Bhattacharyya coefficient is used to calculate the similarity between two probability distributions, assuming that the two densities are $P_1(x)$, $P_2(x)$. Then, the BC of $P_1(x)$ and $P_2(x)$ in the continuous domain is defined as follow.

$$BC(P_1, P_2) = \int \sqrt{P_1(x)P_2(x)}dx \qquad (3.1)$$

The BC of $P_1(x)$ and $P_2(x)$ in the discrete domain is defined as follow.

$$BC(P_1, P_2) = \sum_{x \in X} \sqrt{P_1(x)P_2(x)} \qquad (3.2)$$

In the recommendation system, suppose that $P_1(x)$ and $P_2(x)$ are the rating data of user u and user v, respectively, then the BC similarity between user u and user v is shown in Eq. (3.3).

$$sim_{BC}(u, v) = BC(\widehat{P_u}, \widehat{P_v}) = \sum_{i=1}^{N} \sqrt{\left(\widehat{P_{ui}}\right)\left(\widehat{P_{vi}}\right)} \qquad (3.3)$$

Among them, P_u and P_v are used to evaluate the dispersion of the rating of user u and user v, respectively, and N is the total number of categories of ratings. $\widehat{P_{u\ell}} = num(h)/num(u)$, Num($h$) represents the total number of items that were rated by the user u as h, and num(u) represents the total number of items the user u rated.

The BC similarity between item i and item j is:

$$sim_{BC}(i,j) = BC(\widehat{P}_i, \widehat{P}_j) = \sum_{v=1}^{N} \sqrt{\left(\widehat{P_{iv}}\right)\left(\widehat{P_{jv}}\right)}. \qquad (3.4)$$

P_i and P_j are used to evaluate the dispersion of the rating of item i and item j, respectively. N is the total number of users rated, $\widehat{P_{\ell v}} = num(v)/num(i)$, and num(v) is the total number of users who rated item i as v. Num(i) represents the total number of users who have rated item i.

3.2 LDA Topic Model

LDA is a production three-layer probability model consisting of three parts: document, topic, and word. The so-called production model means that each word in a document is obtained through a two-step process. First, a topic is selected from the distribution of the document topic with a certain probability, and then a word is selected with a certain probability from the word distribution corresponding to the topic. The probability of occurrence of each word in a document can be expressed as follow.

$$p(\text{word} \mid \text{document}) = \sum_{\text{topic}} p(\text{word} \mid \text{topic}) \times p(\text{topic} \mid \text{document}) \qquad (3.5)$$

For the m-th document in data set D, M is the number of documents, and the LDA probability map model can be obtained, as shown in Fig. 1.

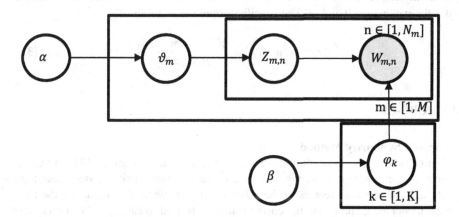

Fig. 1. LDA probability map model

As can be seen from Fig. 1, LDA is a three-layer model, φ_k represents the probability distribution of words in topic k, and ϑ_m represents the topic probability distribution of the m-th document. K represents the number of topics, and α and β are parameters of the Dirichlet distribution, which are used to generate the topic probability distribution ϑ_m corresponding to the document m and the word probability distribution

φ_k corresponding to the topic k. The Dirichlet distribution is a probability metric function defined in the real interval [0, 1], which is expressed as follows.

$$Dir(\theta \mid \alpha) = \frac{\Gamma(\sum\limits_{k=1}^{K} \alpha_k)}{\prod\limits_{k=1}^{K} \Gamma(\alpha_k)} \prod\limits_{k=1}^{K} \theta_k^{\alpha_k-1} \tag{3.6}$$

When the parameters α and β have been given, the joint probability distribution of a document in the LDA topic model is shown as follows.

$$p(\theta, \varphi, \mathbf{z}, \mathbf{w} \mid \alpha, \beta) = p(\varphi \mid \beta) \prod_{n=1}^{N} p(w_n \mid \varphi_{zn}) p(z_n \mid \theta) p(\theta \mid \alpha) \tag{3.7}$$

Assume that the number of documents in the corpus Ω is M. There are N words in each document. By integrating θ and summing z, the available probability distribution of Ω is shown as follows.

$$p(\Omega, \mid \alpha, \beta) = \prod_{d=1}^{M} \int p(\varphi \mid \beta)(\prod_{n=1}^{N} \sum_{z_{d,n}} p(w_{d,n} \mid \varphi_{z_{d,n}}) p(z_{d,n} \mid \theta_d) p(\theta_d \mid \alpha)) d\theta_d \tag{3.8}$$

There are two important parameters in the LDA topic model that need to be estimated. They are the document-theme probability distribution ϑ_m and the topic-word probability distribution φ_k. In this paper we use ϑ_m. We use the posterior probability to infer the parameters θ and φ. The specific formula is as follows:

$$p(\mathbf{z}, \theta, \varphi \mid w, \alpha, \beta) = \frac{p(\theta, \varphi, \mathbf{z}, \mathbf{w} \mid \alpha, \beta)}{p(\Omega \mid \alpha, \beta)} \tag{3.9}$$

3.3   UBL-CF

Improved Similarity Method
When calculating the topic similarity, we use the method of literature [23] to treat all user and item corresponding tag sets as a corpus. Assume that the document-topic probability distribution vectors for item i and item j obtained after modeling the LDA theme model are t_i and t_j, and the cosine similarity is used to calculate the similarity of the document-topic probability distribution between item i and item j, the formula is shown as Eq. (3.10).

$$sim(i,j) = \cos(\mathbf{t}_i, \mathbf{t}_j) = \frac{\mathbf{t}_i \cdot \mathbf{t}_j}{\|\mathbf{t}_i\| \cdot \|\mathbf{t}_j\|} \tag{3.10}$$

$$sim_{rating}(u,v) = \frac{\sum_{j \in I_{uv}} (R_{uj} - \overline{R_u})(R_{vj} - \overline{R_v})}{\sqrt{\sum_{j \in I_{uv}} (R_{uj} - \overline{R_u})^2}\sqrt{\sum_{j \in I_{uv}} (R_{vj} - \overline{R_v})^2}} \qquad (3.11)$$

$$sim_{rating}(i,j) = \frac{\sum_{u \in U_{ij}} (R_{ui} - \overline{R_i})(R_{vj} - \overline{R_j})}{\sqrt{\sum_{u \in U_{ij}} (R_{ui} - \overline{R_i})^2}\sqrt{\sum_{u \in U_{ij}} (R_{vj} - \overline{R_j})^2}} \qquad (3.12)$$

We use the Pearson Correlation Coefficient to calculate the rating similarity of users or items. The formulas are shown as Eqs. (3.11) and (3.12). Assume that the topic similarity calculated between user u and user v is expressed as $sim_{LDA}(u,v)$, and the rating similarity is expressed as $sim_{rating}(u,v)$. The topic similarity between item i and item j is denoted as $sim_{LDA}(i,j)$, and the rating similarity is represented as $sim_{rating}(i,j)$. Then, the improved user local similarity $sim_{Loc}(u,v)$ and the item local similarity $sim_{Loc}(i,j)$ are respectively expressed as follows.

$$sim_{Loc}(u,v) = \alpha\, sim_{LDA}(u,v) + (1 - \alpha)sim_{rating}(u,v) \qquad (3.13)$$

$$sim_{Loc}(i,j) = \beta\, sim_{LDA}(i,j) + (1 - \beta)sim_{rating}(i,j) \qquad (3.14)$$

Among them, the parameters α and β are balance parameters which controls the proportion of rating similarity and similarity of the tag theme. After combining the similarities of the Bhattacharyya coefficients, the improved similarities of users or items are respectively expressed follows.

$$sim_{unify}(u,v) = sim_{BC}(u,v)sim_{Loc}(u,v) \qquad (3.15)$$

$$sim_{unify}(i,j) = sim_{BC}(i,j)sim_{Loc}(i,j) \qquad (3.16)$$

Predict Rating

This paper proposes three improved collaborative filtering algorithms, namely UBL-UserCF algorithm, UBL-ItemCF algorithm, UBL-HybridCF algorithm.

UBL-UserCF

The equation that use UBL-UserCF to predict the user u's rating for item i is shown as follows.

$$R_{ui} = \overline{R_u} + \frac{\sum_{v \in N(u)} sim_{unify}(u,v)(R_{vi} - \overline{R_v})}{\sum_{v \in N(u)} sim_{unify}(u,v)} \qquad (3.17)$$

$sim_{unify}(u,v)$ is the similarity calculated by Eq. (3.15).

UBL-ItemCF

The equation that use UBL-ItemCF to predict the user u's rating for item i is shown as follows.

$$R_{ui} = \overline{R}_i + \frac{\sum\limits_{j \in N(i)} sim_{unify}(i,j)(R_{uj} - \overline{R}_j)}{\sum\limits_{j \in N(i)} sim_{unify}(i,j)} \tag{3.18}$$

$sim_{unify}(i,j)$ is the improved similarity calculated by Eq. (3.16).

UBL–HybridCF

The equation that use UBL – HybridCF to predict the user u's rating for item i is shown as follows.

$$\widehat{R} = \lambda \times \left(\overline{R}_u + \frac{\sum\limits_{v \in N(u)} sim_{unify}(u,v)(R_{vi} - \overline{R}_v)}{\sum\limits_{v \in N(u)} sim_{unify}(u,v)} \right) + (1 - \lambda)$$

$$\times \left(\overline{R}_i + \frac{\sum\limits_{j \in N(i)} sim_{unify}(i,j)(R_{uj} - \overline{R}_j)}{\sum\limits_{j \in N(i)} sim_{unify}(i,j)} \right) \tag{3.19}$$

The left term is improved user-based collaborative filtering (UBL-UserCF) and the right term is improved item-based collaborative filtering (UBL-ItemCF). Balanced parameter λ determines the proportion of the two algorithms.

4 Experiments

4.1 Experimental Data

In this experiment, we use the ML-100 K data set, which includes 943 users, 1,682 movies, and 100,000 rating data. In this dataset, each user rated at least 20 movies. The rating scale of the movie is represented by five integers (1, 2, 3, 4, 5). The level of the rating describes the user's preference for the movie. The higher the rating, the more interested the user is in the movie.

4.2 Evaluation Metric

The Mean Absolute Error (MAE) calculates the prediction error using the absolute value of the difference between the real score and the predicted score, because its formula is easy to understand and simple to calculate, and it has been widely used in the evaluation of the recommendation system. The lower the MAE value is, the higher the accuracy of recommendation system recommendation is.

Assume that the predicted ratings are $\{P_{u1}, P_{u2}, P_{u3}, \dots, P_{uN}\}$ and the real ratings are $\{r_{u1}, r_{u2}, r_{u3}, \dots, r_{uN}\}$, the MAE is:

$$MAE = \frac{\sum_{i \in N} |P_{ui} - r_{ui}|}{N} \tag{4.1}$$

4.3 Experimental Results and Analysis

In order to verify the superiority of UBL-CF algorithm proposed in this paper, we compare our methods with the existing JMSD algorithm [25], UserCF algorithm [6], ItemCF algorithm [7], HybridCF algorithm [12], and ULR-CF algorithm [23].

When using the LDA topic model to mine topic similarity, the number of topics is set to 6 based on empirical values. For the UBL-UserCF algorithm, it has been found through experiments that the parameter α in Eq. (3.13) has the best effect with 0.6; for the UBL-ItemCF algorithm, the optimal value of parameter β in Eq. (3.14) is set to 0.2; For the UBL-HybridCF algorithm, the optimal value of parameter λ in Eq. (3.19) is 0.2. The detailed analysis is shown as follows.

Comparison Between UBL-UserCF and Other Three Algorithms
We set the size of the neighbor set K from 10 to 50 in steps of 5. Figure 2 shows the experimental results.

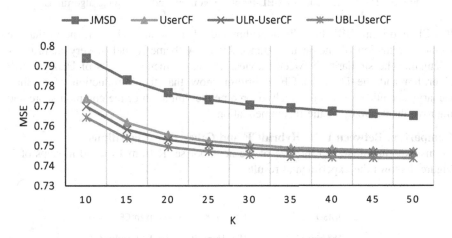

Fig. 2. The MAE values of UBL-UserCF algorithm and other three algorithms.

From Fig. 2, it can be found that the four algorithms show a decreasing trend with the increase of the value of K, indicating that the more neighbors referenced in the experiment, the more accurate the recommendation results can be obtained. The prediction error of the UBL-UserCF algorithm proposed in this paper is obviously less than the other three algorithms, which shows that the introduction of the Bhattacharyya coefficient and the LDA theme model can effectively improve the quality of the recommendation.

Comparison Between UBL-ItemCF and Other Three Algorithms
In this experiment, we set the size of the neighbor set K from 10 to 50 in steps of 5. Figure 3 shows the experimental results.

As shown in Fig. 3, the MAE curve of UBL-ItemCF algorithm is significantly lower than other algorithms, which shows that its prediction error is smaller and the recommendation result is more accurate. Comparing the experimental results of ULR-

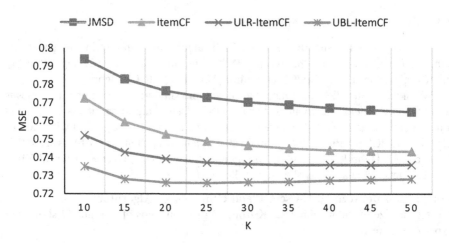

Fig. 3. The MAE values of UBL-ItemCF algorithm and other three algorithms.

ItemCF algorithm, UBL-ItemCF algorithm and other two algorithms, it shows that the theme information of movie tags mined by LDA theme model is very helpful for calculating the similarity between movies. The experimental results of UBL-ItemCF algorithm and the ULR-ItemCF algorithm show that the introduction of a Bhattacharyya coefficient can effectively overcome the limitation of the common score and improve the accuracy of the recommendation.

Comparison Between UBL-HybridCF and Other Five Algorithms
In this experiment, we set the size of the neighbor set K from 10 to 50 in steps of 5. Figure 4 shows the experimental results.

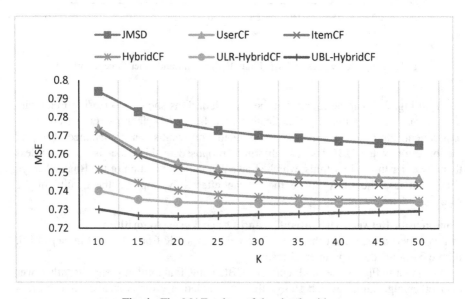

Fig. 4. The MAE values of the six algorithms.

It can be clearly seen from Fig. 4 that the algorithm proposed in this paper has a smaller MAE value, that is, the prediction error is smaller. Through three groups of comparison experiments, the validity of the UBL-CF algorithm proposed in this paper was verified.

5 Conclusion

In this paper, we propose an improved collaborative filtering algorithm unified Bhattacharyya coefficient and LDA topic model (UBL-CF). UBL-CF use the LDA topic model to mine potential topic information in the tag set and embed the underlying topic information into the calculation of similarity. At the same time, the introduction of the Bhattacharyya coefficient was used to alleviate the data sparsity problem without common scoring. Compared with other collaborative filtering algorithms (the JMSD, the UserCF, the ItemCF, the HybridCF, and ULR-CF), the UBL-CF proposed in this paper has a great improvement in the recommendation performance. In the future, we will continue to research the problems of sparse and cold start in traditional collaborative filtering.

Acknowledgments. This work is supported by National Natural Science Foundation of China under Grant 61432008, 61272222.

References

1. Mennel, P.A., Melgoza, P., Gyeszly, S.D.: Information overload. Collect. Build. **21**(21), 32–43 (2002)
2. Anidorifón, L., Santosgago, J., Caeirorodríguez, M., et al.: Recommender systems. Commun. ACM **40**(3), 56–58 (2015)
3. Schafer, J.B., Konstan, J.A., et al.: E-commerce recommendation applications. Data Min. Knowl. Discov. **5**(1–2), 115–153 (2001)
4. Pazzani, M.J., Billsus, D.: Content-based recommendation systems. In: Brusilovsky, P., Kobsa, A., Nejdl, W. (eds.) The Adaptive Web. LNCS, vol. 4321, pp. 325–341. Springer, Heidelberg (2007). https://doi.org/10.1007/978-3-540-72079-9_10
5. Melville, P., Mooney, R.J., Nagarajan, R.: Content-boosted collaborative filtering for improved recommendations. In: Eighteenth National Conference on Artificial Intelligence, pp. 187–192. American Association for Artificial Intelligence (2002)
6. Resnick, P., Iacovou, N., Suchak, M., et al.: GroupLens: an open architecture for collaborative filtering of netnews. In: ACM Conference on Computer Supported Cooperative Work, pp. 175–186. ACM (1994)
7. Linden, G., Smith, B., York, J.: Amazon.com recommendations: item-to-item collaborative filtering. IEEE Internet Comput. **7**(1), 76–80 (2003)
8. Guo, G., Zhang, J., Yorke-Smith, N.: TrustSVD: collaborative filtering with both the explicit and implicit influence of user trust and of item ratings. In: Twenty-Ninth AAAI Conference on Artificial Intelligence, pp. 123–129. AAAI Press (2015)
9. Liu, H., Hu, Z., Mian, A., et al.: A new user similarity model to improve the accuracy of collaborative filtering. Knowl. Based Syst. **56**(3), 156–166 (2014)

10. Aggarwal, C.C.: Knowledge-based recommender systems. In: Aggarwal, C.C. (ed.) Recommender Systems, pp. 167–197. Springer, Cham (2016). https://doi.org/10.1007/978-3-319-29659-3_5

11. Felfernig, A., Burke, R.: Constraint-based recommender systems: technologies and research issues. In: International Conference on Electronic Commerce, p. 3. ACM (2008)

12. Ji, H., Li, J., Ren, C., et al.: Hybrid collaborative filtering model for improved recommendation. In: IEEE International Conference on Service Operations and Logistics, and Informatics, pp. 142–145. IEEE (2013)

13. Lucas, J.P., Luz, N., Anacleto, R., et al.: A hybrid recommendation approach for a tourism system. Expert Syst. Appl. **40**(9), 3532–3550 (2013)

14. Ji, K., Sun, R., Li, X., et al.: Improving matrix approximation for recommendation via a clustering-based reconstructive method. Neurocomputing **173**(P3), 912–920 (2016)

15. Barua, S., Gao, X., Pasman, H., et al.: Bayesian network based dynamic operational risk assessment. J. Loss Prev. Process Ind. **41**, 399–410 (2016)

16. Xuan, N.L., Vu, T., Le, T.D., et al.: Addressing cold-start problem in recommendation systems. In: International Conference on Ubiquitous Information Management and Communication, pp. 208–211. ACM (2008)

17. Almazro, D., Shahatah, G., Albdulkarim, L., et al.: A survey paper on recommender systems. Comput. Sci. (2010)

18. Sun, D., Li, C., Luo, Z.: A content-enhanced approach for cold-start problem in collaborative filtering. In: International Conference on Artificial Intelligence, Management Science and Electronic Commerce, pp. 4501–4504. IEEE (2011)

19. Wang, W., Zhang, D., Zhou, J.: COBA: a credible and co-clustering filterbot for cold-start recommendations. In: Wang, Y., Li, T. (eds.) Practical Applications of Intelligent Systems. Advances in Intelligent and Soft Computing, vol. 124, pp. 467–476. Springer, Cham (2011). https://doi.org/10.1007/978-3-642-25658-5_56

20. Ge, S., Ge, X.: An SVD-based collaborative filtering approach to alleviate cold-start problems. In: International Conference on Fuzzy Systems and Knowledge Discovery, pp. 1474–1477. IEEE (2012)

21. Wang, Z., Wang, Y., Wu, H.: Tags meet ratings: improving collaborative filtering with tag-based neighborhood method. In: The Workshop on Social Recommender Systems. IUI (2010)

22. Ji, K., Shen, H.: Addressing cold-start: scalable recommendation with tags and keywords. Knowl. Based Syst. **83**(1), 42–50 (2015)

23. Na, G., Yang, M.: Topic model embedded in collaborative filtering recommendation algorithm. Comput. Sci. **43**(3), 57–61 (2016)

24. Zhao, X., Niu, Z., Chen, W., et al.: A hybrid approach of topic model and matrix factorization based on two-step recommendation framework. J. Intell. Inf. Syst. **44**(3), 335–353 (2015)

Author Index

Printed in the United States
By Bookmasters